Soil Micromorphology

VOLUME 2

Soil Genesis

Soil Micromorphology

VOLUME 2
Soil Genesis

Edited by
P Bullock and C P Murphy
Rothamsted

A B ACADEMIC PUBLISHERS

Published in the UK by
A B Academic Publishers
P O Box 97
Berkhamsted
Herts HP4 2PX

British Library Cataloguing in Publication Data:
Soil Micromorphology
Vol. 2: Soil genesis
1. Soils – Analysis
I. Bullock, P.   II. Murphy, C.P.
631.4'1   S593

ISBN 0-907360-07-6

Proceedings of the International Working Meeting on Soil Micromorphology, held in August 1981
In two volumes
(Volume 1: Techniques and Applications; Volume 2: Soil Genesis)
ISBN 0-907360-07-6   Volume 2
© 1983 A B Academic Publishers
Printed in the Netherlands by ICG Printing Dordrecht

Preface

Since the publication of W.L. Kubiena's *Micropedology* in 1938, there has been a major expansion of soil micromorphology both in the number of scientists using it and in the number of disciplines in which it is being applied. This increasing interest is reflected in the decision of the International Society of Soil Science to establish a Sub-Commission of Soil Micromorphology in 1978.

International Working-Meetings on Soil Micromorphology have been held at regular 4–5 years intervals since 1958. The papers contained in these two volumes were delivered at the sixth Working-Meeting held in London in August, 1981. This meeting was the first one of the newly-formed Sub-Commission.

The papers presented at the Working-Meeting were of two types: (1) keynote papers providing a state-of-the-art view of some of the more important areas of soil micromorphology, and (2) poster papers presenting new research material. A selection of the papers presented are included in these volumes. Volume 1 is devoted to 'Techniques and Applications'; in both of these areas there has been considerable development. Volume 2 is devoted to 'Soil Genesis', the area with which, historically, soil micromorphology has been most associated.

The Proceedings of the previous meetings have served as a useful summary of the progress and the direction of soil micromorphology. It is intended that these two Volumes should serve likewise and to some extent compensate for the few textbooks available in this particular area.

All the papers have been refereed by the members of the Organising Committee of the Working-Meeting. Thanks are extended to the following for performing this task and for their general help with the preparation of the Meeting: Dr J.A. Catt (Rothamsted), Dr J.F. Collins (University College, Dublin), Dr J.D. Dalrymple (Reading University), Dr E.A. FitzPatrick (Aberdeen University), Dr D.A. Jenkins (University College of North Wales, Bangor), Dr P.J. Loveland (Soil Survey of England and Wales, Rothamsted), and Mr L. Robertson (Macaulay Institute for Soil Research, Aberdeen). Grateful thanks are also extended to Mr D. Mackney, Head of the Soil Survey of England and Wales, for his support.

P. Bullock
C.P. Murphy
(Editors)

These proceedings are respectfully dedicated to the memory of Dr A. Jongerius (1925–1982) in recognition of his very considerable contribution to soil micromorphology.

Contents

x

VOLUME 1: TECHNIQUES AND APPLICATIONS

CLAY SKINS AND ARGILLIC HORIZONS

J.A. McKeague

Land Resource Research Institute,
Agriculture Canada, Ottawa
     (LRRI Publication No. 113)

ABSTRACT
     Publication in 1960 of the 7th approximation of the new United States' system of soil taxonomy precipitated studies in many countries to test the application of the concept and definition of the argillic horizon on local soils. New knowledge was obtained and numerous problems in application of the concept and criteria of the argillic horizon became evident. This review focuses on problems associated with the argillic horizon and suggests routes to the solution of some of them.
     The concept and definition of the argillic horizon were developed mainly from studies of soils developed in calcareous late Pleistocene and Holocene deposits. For such soils, the argillic horizon is a useful diagnostic criterion at a high categorical level. Improvements are required, however, in the application of argillic horizon criteria in soil taxonomy. For some paleosols and especially for highly weathered soils of tropical regions the concept of the argillic horizon is less meaningful and its use as a diagnostic criterion at high levels is questionable.
     Study of thin sections provides useful evidence of illuvial clay in some soils but interpretations are highly subjective. There is a major need for increased use of reference thin sections and workshops aimed at increasing the reliability and uniformity of descriptions and interpretations in soil micromorphology.

INTRODUCTION
     Translocation of colloidal clay in suspension from surface to subsurface soil horizons was mentioned in the German and Russian soil literature of the early decades of this century (Rode, 1970). It was probably not a major concept of early Russion pedologists as Glinka (1914) referred only casually to the process in

his summary of pioneering Russian pedology. Robinson
(1932) wrote clearly of clay eluviation-illuviation
resulting in texture differentiation in soils but general
acceptance of the process probably occurred only in the
1950s. A horizon of clay illuviation was not a basis
of definition of taxa in early systems of soil classification (Glinka, 1914; Baldwin et al., 1938),
though some of the classes in those systems would
probably correlate well with Alfisols (Soil Survey
Staff, 1975). Eluviation-illuviation of silicate clay
was not clearly differentiated in those systems from
eluviation of weathering products from A horizons and
deposition of new substances in B horizons (Baldwin et al.,
1938; Fridland, 1958).

In the soils literature of the last 50 years, clay
increase from A to B horizons has been attributed to one
or more of the following: discontinuity in texture of
the parent material, translocation of inherited clay,
clay formation by weathering in situ, breakdown of
clay in A horizons and clay synthesis from solution in B
horizons (Brewer and Sleeman, 1970). Micromorphology
was used only rarely before 1950 to aid in resolving the
question of genesis of textural B horizons (Frei and
Cline, 1949; Brewer and Sleeman, 1970).

Since 1960, hundreds of papers have been published
on aspects of characterization and genesis of soils
having B horizons of finer texture than the overlying
A horizons. This work was precipitated, in part, by
publication of the 7th approximation (Soil Survey Staff,
1960). It was fundamental to the system that the
concepts would require modification as knowledge grew
and that the formulation of precise definitions would
provoke their testing. The concept and definition of
the argillic horizon have been tested and knowledge of
such horizons has increased greatly, in part due to
widespread application of micromorphology. Changes
have been proposed in definitions of the argillic
horizon and in its use as a diagnostic horizon in soil
taxonomy but no concensus has emerged on the changes
required.

The goals of this review are to outline the current
state-of-the-art on concepts, definition and significance
of the argillic horizon and to consider improvements,
including improved uniformity of application of the
definition in classifying soils. Though much literature
related to argillic horizons was read, the treatment of
the subject is undoubtedly biased due to the "tint of
the author's lenses" and the unavoidable "selective
focusing" involved in appraising the literature
(Simonson, 1978). The selection of cited literature
favours recent publications available in English.
Like the definition of the argillic horizon (Soil Survey
Staff, 1975) this paper will serve a useful purpose if
it provokes discussion and more perceptive treatments
of the subject.

CLAY MOBILISATION AND DEPOSITION
    Evidence from detailed studies of a wide variety of
soils and from laboratory experiments on clay translocation has been summarised and interpreted in several
publications (Brewer and Sleeman, 1970; Soil Survey
Staff, 1975; De Coninck et al., 1976; Duchaufour,
1977; Gile and Grossman, 1979; Eswaran, 1979). The
general process of clay mobilisation and deposition
seems to occur as follows. When rain falls upon dry
soil, some fine material may be dispersed and it may
remain in suspension as the water moves downward
through major voids. If the subsoil is dry, the downward moving water will be absorbed into the soil mass
and the suspended clay will be filtered out and deposited
on the walls of the voids.
    Many complexities can be added to this simple
conceptual model of clay translocation. Factors that
favour clay translocation are: alternating wet and
dry conditions, a system of macrovoids in the soil,
absence of cements such as sesquioxides and carbonates,
pH between about 4.5 and 6.5 (low exchangeable Al and no
excess Ca and Mg), or high pH associated with exchangeable
Na, and zero point of net charge different from soil pH
(Eswaran, 1979). Soluble organic matter may contribute
to the peptization of clay, and leaching of salts and
carbonates may leave voids that favour local movement

of clay and more rapid flow of soil water. The
deposition of clay moving downward in suspension is
probably favoured by dry subsoil, high pH perhaps due to
carbonates, and to less porous material that slows the
rate of flow.

Though fine clay may be translocated more readily
than coarse clay (Soil Survey Staff, 1975) and though
certain clay minerals may be mobilised preferentially
(De Coninck et al., 1976), coarse clay of varied
mineralogy is translocated in soils (Eswaran, 1979).

CONCEPT AND DEFINITION OF THE ARGILLIC HORIZONS

"An argillic horizon is an illuvial horizon in which
layer-lattice silicate clays have accumulated by
illuviation to a significant extent". (Soil Survey Staff,
1975). This brief definition embodies the concept but
specific criteria are necessary to foster uniformity in
recognition of argillic horizons. The specific
criteria are:

1. More clay by specified amounts than the overlying
    eluvial horizon, and normally a higher ratio of fine
    to total clay than the underlying horizon.
2. It must exceed a certain thickness (7.5 to 15 cm).
3. If structureless, it must have oriented clay
    bridging sand grains and in some pores.
4. If pedal, it must have clay skins on ped surfaces or
    have oriented clay occupying 1% or more of the cross
    section. This requirement is waived if the illuvial
    horizon is clayey with 2-to-1 lattice clays, if there
    is evidence of swelling and if other evidence
    indicates clay eluviation-illuviation.
5. If there is a lithologic discontinuity or a plough
    layer immediately above the argillic horizon, the
    horizon must have clay skins in pores or on ped sur-
    faces. It should have oriented clay bodies occupying
    1% or more of the cross section or the highest ratio
    of fine to total clay.

Establishment of the diagnostic argillic horizon
was probably based on several underlying concepts. One
of these is that accessory properties of the horizon
are useful in understanding soils and interpreting their
behavior (Soil Survey Staff, 1975; Arnold, 1979).
Another is that coatings of clay on ped surfaces affect

water and nutrient movement and plant growth in a
different way than clay dispersed throughout the soil
matrix (Khalifa and Buol, 1968; Miller et al., 1971).
     Though the specific criteria of the argillic
horizon are the result of a concerted effort to define
measureable properties that reflect the results of the
genetic process of clay translocation, they have not
led to unambiguous recognition of argillic horizons
(Bronger, 1978; Arnold, 1979; Isbell, 1980a;
McKeague et al., 1981). The procedures involved in
determining the diagnostic properties are not adequately
standardised (Isbell, 1980b; McKeague et al., 1980).
Perhaps more fundamental is the question of whether this
set of diagnostic criteria sufficiently demark argillic
horizons from all that do not fit the concept (Arnold,
1979; Eswaran, 1979).

PROBLEMS IN APPLYING THE CRITERIA OF ARGILLIC HORIZONS
     Response to a questionnaire directed to pedologists
outside the United States showed that application of
diagnostic features of argillic horizons was a major
problem in using Soil Taxonomy (Cline, 1980). In this
section each of the operations involved in identifying
an argillic horizon is considered; emphasis is given to
clay skins and to the recognition of illuvial clay in
thin section.

Clay Increase
     In many soils with well developed argillic horizons,
particle-size analysis is hardly necessary to confirm
the clay increase. Such soils may have a light coloured,
sandy to loamy eluvial horizon overlying a clayey,
blocky-structured B horizon, with obvious clay skins
on the ped surfaces, that tongues into an underlying
calcareous horizon. In other soils, however, particle-
size analysis is essential to check the clay increase
from A to B. Though particle-size data are method-
dependent, comparisons of particle-size classes of
associated A and B horizon samples in a reliable
laboratory are probably adequate. The reliability of
determination of fine clay is more questionable.
Furthermore, the ratio of fine to total clay commonly
reaches a maximum in a horizon other than the argillic

horizon (Beinroth and Panichapong, 1979; Isbell,
1980a), though such a maximum is common in argillic
horizons of young soils (Smith and Wilding, 1972;
Soil Survey Staff, 1975).

Clay Skins
     Consistent application of argillic horizon criteria
depends to a major extent on the ability of pedologists to
recognise clay skins on ped or grain surfaces and lining
voids. In spite of this, there have been few studies on
consistency of identification of clay skins and no
thoroughly-documented published guidelines on how to
recognise clay skins indicative of clay illuviation.
Problems in identification of clay skins are well
documented, both for weathered soils near the Oxisol-
Ultisol boundary (Smith et al., 1975; Moorman, 1978;
Eswaran, 1979; Buol, 1980; Isbell, 1980a and b) and
for some weakly weathered soils of temperate areas
(McKeague et al., 1981; Smeck et al., 1981). In the
latter areas, the problem seems to be greatest in some
poorly drained soils in which bleached ped faces may
appear to be clay skins.
     Consistency of recognition of clay skins could be
improved by developing guidelines for identifying them
and by comparisons of macro-observations with descriptions
of thin sections. The observer should examine fresh ped
surfaces and intraped voids with a 10 to 20 power hand
lens or, preferably, a stereomicroscope when the
material is moist to dry, and compare ped faces with ped
interiors. If clay skins are present on ped surfaces,
protruding sand grains will be covered by a smooth
coating. Commonly clay skins have a waxy appearance
and display flow lines (Soil Survey Staff, 1975).
Problems will remain, however, especially with clayey
and highly weathered soils, even with careful and
consistent application of guidelines.
     The fact that the most obvious clay skins commonly
occur in horizons well below the horizon of maximum clay
content may be thought of as a problem in logic of
application of argillic horizon criteria. In the
discussion of the argillic horizon in Soil Taxonomy
(Soil Survey Staff, 1975) however, it is recognised that
argillans in the upper part of the horizon are subject

to disruption by various processes and that argillans
are more stable in deeper horizons. If the evidence
indicates uniform material, argillans in the lower B
horizons are useful evidence that a horizon of maximum
clay probably is an argillic horizon.

Oriented Clay in Thin Section
        Application of this criterion of argillic horizon is
fraught with many problems. Though the definition
states, "oriented clay in one percent or more of the
cross section" (Soil Survey Staff, 1975) the intention
is oriented clay that has been translocated and
deposited. Much of the oriented clay in some soils is
in situ, inherited clay and some is clay formed by
weathering. This may be a minor ambiguity in the
definition but it has caused some confusion.
        A more serious problem is that of judging whether
a specific body of oriented clay in thin section is
illuvial clay. Guidelines have been suggested by many,
and the following incorporate most of those listed by
Brewer (1964), Hill (1970), Bullock and Murphy (1979):
on a ped surface or lining a void and strongly oriented
parallel to the surface; microstratification indicating
successive deposition; sharp boundary with the s-matrix
visible under plane light and between crossed
polarisers; different in colour, orientation, uniformity
or other features from any inherited bodies of oriented
clay that may occur in the C horizon. No set of
guidelines is adequate, however, and Brewer and Sleeman
(1970) have pointed out the need to study the overall
properties of the pedon including thin sections of all
horizons before making interpretations.
        Some of the problems that arise are illustrated,
but first an ideal example of an argillic horizon of
a Cryoboralf is shown (Fig. 1). The argillans coat
both sides of the planar void continuously, they are
strongly oriented, show microlamination, are free of
sand grains, and are clearly different from the matrix
which contains no bodies of oriented clay. Unfortunately,
micromorphological evidence of illuvial clay is not always

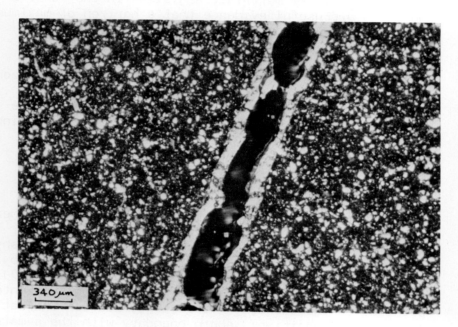

Fig. 1.   Strongly oriented argillan on both walls
          of a planar void in a Cryoboralf. XPL.

so obvious. Many clayey soils have stress argillans
but usually they are not sharply differentiated from
the matrix (Fig. 2).   In some sandy soils, most of the
fine matrix material may occur as coatings on skeleton
grains (Fig. 3a) and the coatings may be oriented,
perhaps due to wetting and drying cycles (Nettleton
et al., 1969).   Such cutans may be inherited from the
parent rock (Collins et al., 1978), formed by local
re-arrangement of the fabric, or be constituted, in
part, of illuvial clay (Wieder and Yaalon, 1978;
Wang and McKeague, 1982).   Planar void argillans
in a B horizon may be similar to inherited, elongated,
oriented bodies of clay in the C horizon (Fig. 3b).
Such argillans could result from the development of

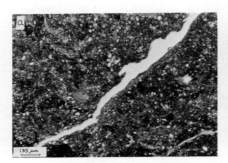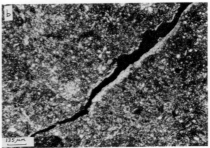

Fig. 2. Stress argillans on the walls of a planar
         void in the cambic horizon of a
         Cryochrept;   (a) In PPL and (b) In XPL.

longitudinal cracks through the inherited clay bodies
or from illuviation. Oriented clay bodies in
horizons depleted of carbonates may be due to local
accumulation of clay contained in the carbonate-rich
material or to illuviation of clay from above. Illuvial
clay may be deposited with silt and sand and occur as
heterogeneous, weakly oriented cutans especially in
subsoils of cultivated soils (Kwaad and Mücher, 1979).
Silty cutans that are apparently illuvial also occur
in some uncultivated soils (Fig. 3c). Void argillans
and papules of similar appearance may be local
weathering products of biotite or other minerals
(Mermut and Jongerius, 1980). The problem of
distinguishing illuvial clay from clay formed in place
by weathering probably is most difficult with paleosols.
     The problem of estimating the proportion of a
horizon occupied by illuvial oriented clay is almost
as complex as that of deciding whether a specific
body of clay is illuvial. One aspect of the problem is
the variability of distribution of illuvial clay and
hence the need to study many thin sections and count
many points in each (Brewer, 1964; Milfred et al.,
1967). Another aspect is the differences that result
from method of counting; results of point counting
are approximately double those of grid counting
(McKeague et al., 1980). Problems of recognition of

J.A. McKeague

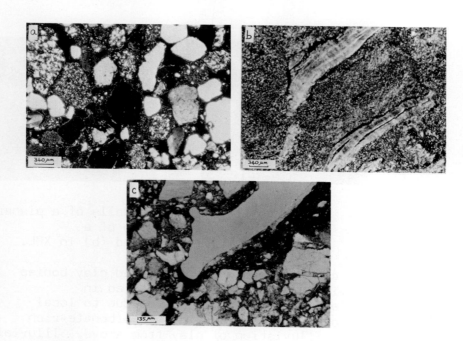

Fig. 3. (a) Weakly oriented clayey cutans on sand
         grains in the cambic horizon of a Cryochrept;
         (b) Elongated, oriented clay bodies in the C
         horizon of a Cryoboroll; (c) Compound cutan,
         argillan and thick siltan, on the bottom of a
         vugh in the cambic horizon of a Dystrochrept.
         (a and b both XPL, c in PPL).

illuvial clay and of methods of estimation are combined
in comparisons of estimates of illuvial clay in the same
thin sections by different operators (McKeague et al.,
1980). The results (Table 1) demonstrate the urgent
need for reference thin sections, small working meetings
around the microscope, and the development of clearer
standards on the estimation of illuvial clay. The
question of whether 1% is a suitable minimum amount of
illuvial oriented clay is academic unless consistent
estimates can be achieved.

Table 1. Estimates of illuvial clay in the same thin
   sections by ten operators (McKeague et al., 1980).

Section no.	Range %	Mean %	Standard deviation
1	1.1-4.6	3.2	1.25
2	0.6-2.7*	1.8	0.94
3	0.4-2.1	1.1	0.68
4	1.2-8.3	4.2	2.12
5	0.6-3.8	2.1	1.17
6	1.7-17.8	8.5	5.42

   * One value, 9.4%, was not included.

A POINT OF VIEW ON THE ARGILLIC HORIZON
      Though clear definition and precise criteria that
can be applied unambiguously are worthy goals for argillic
and other diagnostic horizons they are less fundamental
than the ultimate goals of soil taxonomy. These goals
have been outlined in many publications (Smith, 1963;
Soil Survey Staff, 1975). A principle common to most
is that taxonomy should result in soil classes that make
sense in relation to the overall understanding of the
genesis and properties of soils. For mid to late
Pleistocene and Holocene soils developed in calcareous
materials under temperate, humid or subhumid climates,
soil classes based on the argillic horizon make sense.
The sequence of processes from leaching of salts and
carbonates, translocation and deposition of clay and
thus development of eluvial and illuvial horizons,
tongueing of the eluvial horizon into the illuvial
horizon, is well documented (Frei and Cline, 1949;
Smeck et al., 1968; Fedoroff, 1969; Jamagne, 1972).
The concept of the argillic horizon was based mainly
on studies of such soils (Arnold, 1979).
      In recent years, on the other hand, many studies
of older, more highly weathered soils of subtropical

and tropical regions have indicated that evidence of
illuvial clay is commonly ambiguous and that the
presence of an argillic horizon is not one of the key
properties that reflects the important aspects of
genesis and behaviour of these soils (Isbell, 1980a
and b; papers in Beinroth and Panichapong, 1979).
Properties associated with weathering and with the
dominance of low activity clays seem to be much more
important than those associated with minor amounts
of illuvial clay.

        Recent work on soils with low activity clays has
contributed greatly to the knowledge of their
properties and it will probably lead to revisions
in soil taxonomy. The revisions might include changes
in the definition of the argillic horizon to exclude
highly weathered soils or changes in the categorical
level at which an argillic horizon is diagnostic in
such soils. A somewhat analogous case in Canada is
the change in taxonomy of soils having permafrost close
to the surface (Canada Soil Survey Committee, 1978).
Studies of these soils in the last two decades showed
that processes and properties related to freezing and
permafrost were of far greater genetic and practical
significance than those associated with gleying
and oxidation. A new order, Cryosolic, was formed
to classify these soils and it makes sense in relation
to current knowledge, concepts and interpretations.

        The concept and definition of the argillic horizon
has been tested in other old soils and paleosols
(Nettleton et al., 1975; Brinkman et al., 1977;
Bullock and Murphy, 1979; Chittleborough and Oades,
1979; Torrent et al., 1980). Such soils may have
undergone several changes in climate and argillans
may be impregnated with carbonates (Gile and Grossman,
1979), or iron oxides (Bullock and Murphy, 1979), or they
may be partly destroyed by weathering (Brinkman et al.,
1977). The paleo-argillic horizon (Avery, 1980) may
win general acceptance as a type of argillic horizon.
Whether it should be used diagnostically in taxonomy
in the same way as other argillic horizons is an open
question.

Evidence of movement of clay with silt and sand
in both cultivated (Kwaad and Mücher, 1979) and uncultivated (Fedoroff, 1974) soils focuses attention on
the need to establish criteria for taking into account
heterogeneous cutans in assessing clay illuviation.
In the case of 'agricutans' (Kwaad and Mücher, 1979) they
do not indicate a stable surface which is one of the
useful accessory properties of argillic horizons (Soil
Survey Staff, 1975).

RESEARCH NEEDED
     Though much work on clay skins and argillic
horizons has been published during the last two decades,
the need for further research is great. Some of the
major requirements are listed:
1. Improved standardisation of soil micromorphology,
especially quantitative micromorphology (micro-
morphometry). To date, concern about statistically
sound replication of samples (Milfred et al., 1967) and
number of points counted (Brewer, 1964) seems to have
outweighed concern about operator variability (McKeague
et al., 1980). Variability among operators in
estimating illuvial clay in the same thin sections were
astonishing (Table 1). The main reason for the
variability was different concepts of individuals
regarding the features of oriented clay bodies that
constituted adequate evidence of illuvial origin. Other
reasons for differences were method of counting, and
perhaps experience of individuals in soil micromorphology.
The 1% oriented (illuvial) clay criterion for argillic
horizons is of little use unless reasonably uniform
estimates can be made by different operators. The same
applies to all other quantitative estimates in soil
micromorphology.
     Stoops (1978) stressed that micromorphometry should
not be attempted for all soil features seen in thin
section. The fact remains,however,that quantification
is a basic aspect of science. In the words of Lord
Kelvin, "When you can measure that of which you speak
and express it in numbers, you know something about it".
Presumably the quotation implies numbers that other
experienced measurers can duplicate independently.

The need is obvious for reference thin sections and for workshops around the microscope to compare micromorphological observations and interpretations thereof, and to develop standards for identification of illuvial clay and other features of interest. Such attempts at standardisation would be a worthy task for the International Sub Commission on Soil Micromorphology.
2.   Improved standardisation of methods for identifying evidence of illuvial clay in the field. Experienced pedologists do not always agree on whether a particular horizon has clay skins on ped surfaces or lining voids (McKeague et al., 1981). Strongly held concepts of the genesis and appropriate taxonomy of a soil can lead to the acceptance on faith of clay skins somewhere in the control section (Buol, 1980). Field estimates of illuvial clay could be improved by field workshops and by comparisons of field evidence of illuvial clay with micromorphological and other evidence.
3.   Continued work on refining the criteria of an argillic horizon so as to avoid ambiguity (Isbell, 1980a) and capture the physical embodiment of the concept. The usefulness of fine clay to total clay ratios, including the reliability of fine clay data, needs further consideration. In addition, the waiving of the clay skin requirement for clayey argillic horizons having COLE values more than 4% needs further investigation as some such horizons have strongly oriented argillans (Smith and Wilding, 1972; McKeague et al., 1981). Another aspect of this work is to evaluate the levels at which an argillic horizon should be diagnostic in the taxonomy of different soils.
4.   More work on experimental pedology (Hallsworth and Crawford, 1965) would probably be useful in providing improved understanding of processes involved in clay mobilisation and deposition, and formation and destruction of argillans. Previous experiments of this kind have been fruitful (Brewer and Sleeman, 1970; Gombeer, 1977; Nettleton et al., 1969). Some aspects that could be checked are: the possible destruction of illuviation argillans of clayey soils subjected to wetting and drying or freezing and thawing; the role of organic matter in mobilising clay; the significance of translocation and precipitation of Si-Al complexes

(Farmer et al., 1980) in the development of clay-
enriched horizons in some soils, a concept somewhat
similar to that on neoformations that persists in
the Russian literature (Parfenova et al., 1964).
5. Further comprehensive studies are required of
properties of soils from regions with different climates
and different parent materials. Such studies should
include:the geological setting; thorough macro- and
micromorphological description; physical, chemical
and mineralogical data necessary to assess uniformity
of the original material, and extent of transformations
and translocation of constituents; and interpretation
of the integrated information. Such studies involving
specialists in several disciplines and approaches
are more likely to lead to new knowledge and insights
than studies of a single aspect of soil development.
6. Extension of soil micromorphology by use of a variety
of techniques of sub-microscopy will provide specific
information on such aspects as the range of size and
orientation of particles within argillans, and the
composition of weathered argillans (Bisdom and
Jongerius, 1978).

CONCLUSIONS
        The main conclusions from this review are:
1. The concept of the argillic horizon as applied to the
taxonomy of soils developed in Holocene and mid to late
Pleistocene deposits is useful. Further work is needed,
however, to improve the details of argillic horizon
criteria and the uniformity of application of these
criteria in soil taxonomy. In particular, workshops
and use of reference thin sections are required to improve
the uniformity of estimates of illuvial clay in thin
section.
2. In old soils, especially highly weathered soils of
subtropical and tropical regions, the concept of the
argillic horizon and its use as a diagnostic horizon at
high categorical levels of taxonomy are questionable.
The occurrence of minor amounts of identifiable illuvial
clay is of much less importance than other properties of
soils dominated by low activity clays and close to the
present Ultisol-Oxisol boundary. Probably an argillic
horizon should be diagnostic only at the subgroup

level, if at all, in such soils.
3. In paleosols, the argillic horizon may be a useful
indicator of former climates and processes of soil
development, but it may be unrelated to current and
recent pedogenesis. The level at which such paleo-
argillic horizons should be diagnostic in soil taxonomy
is not obvious.

ACKNOWLEDGEMENTS

    I am grateful to the following individuals for
sending recent information and comments on clay skins
and argillic horizons: B.W. Avery, R. Brewer, S.W. Buol,
B. Clayden, P. Bullock, C.P. Murphy, J.R. Sleeman, R.F.
Isbell, R. Miedema, F. De Coninck, A. Bronger, J. Benayas,
L.P. Wilding, and W.D. Nettleton. The photographs were
taken by R.K. Guertin.

REFERENCES

Arnold, R.W. 1979. Concept of the argillic horizon and
    problems in its identification. In : F.H. Beinroth
    and S. Panichapong (Ed), Proceedings Second
    International Soil Classification Workshop. Part II.
    Soil Survey Division, Land Development Department,
    Bangkok, Thailand, 21-33.
Avery, B.W. 1980. Soil Classification for England and
    Wales. Soil Surv. Tech. Monogr. No.14, Harpenden,
    England.
Baldwin, M., Kellogg, C.E. and Thorp, J. 1938. Soil
    classification. In : Soils and Man. U.S.D.A.,
    979-1001.
Beinroth, F.H. and Panichapong, S. (Ed). 1979.
    Proceedings of Second International Soil Classifica-
    tion Workshop. Part II. Soil Survey Division
    Land Development Department, Bangkok, Thailand.
Bisdom, E.B.A. and Jongerius, A. 1978. SEM-EDXRA
    and/or IMMA analysis of cutans, an indurated
    horizon and clayified roots in thin sections of
    some Dutch soils. In : M. Delgado (Ed), Micro-
    morfologia de Suelos. Universidad de Granada,
    Spain, 741-756.

Brewer, R. 1964. Fabric and Mineral Analysis of Soils.
    Wiley, New York. 470 pp.
Brewer, R. and Sleeman, J.R. 1970. Some trends in
    pedology. Earth Sci. Rev., 6, 297-335.
Brinkman, R., Jongmans, A.G. and Miedema, R. 1977.
    Problem hydromorphic soils in north-east
    Thailand. 1. Environment and soil morphology.
    Neth. J. Agric. Sci., 25, 108-125.
Bronger, A. 1978. Climatic sequences of steppe soils
    from eastern Europe and the U.S.A. with emphasis
    on the genesis of the argillic horizon. Catena,
    5, 33-51.
Bullock, P. and Murphy, C.P. 1979. Evolution of a paleo-
    argillic brown earth (Paleudalf) from Oxfordshire,
    England. Geoderma, 22, 225-252.
Buol, S.W. 1980. Morphological characteristics of
    Alfisols and Ultisols. In : B.K.G. Theng (Ed),
    Soils With Variable Charge. New Zealand Soc. Soil
    Sci., 3-16.
Canada Soil Survey Committee. 1978. The Canadian
    System of Soil Classification. Can. Dept. Agric.
    Publ. 1646.   Supply and Services Canada, Ottawa.
Chittleborough, D.J. and Oades, J.M. 1979. The
    development of a red-brown earth. I.A reinter-
    pretation of published data. Aust. J. Soil Res.,
    17, 371-381.
Cline, M.G. 1980. Experience with Soil Taxonomy of the
    United States. Adv. Agron., 23, 193-226.
Collins, J.F., Mullen, G.J. and Kelly, J. 1978. Inherited
    microfabrics of soils derived from micaceous sand-
    stone in Ireland. In : M. Delgado (Ed), Micro-
    morfologia de Suelos. Volume 2. Universidad de
    Granada, Spain, 757-777.
De Coninck, F., Favrot, J.C., Tavernier, R. and Jamagne,
    M. 1976. Degradation dans les sols lessivés
    hydromorphes sur matériaux argilo-sableux: Exemple
    des sols de la nappe détritique Bourbonnaise (France).
    Pédologie, 26, 105-151.
Duchaufour, Ph.1977. Pédologie,   Masson, Paris.

Eswaran, H. 1979. Micromorphology of Alfisols and
    Ultisols with low activity clays. In : F.H.
    Beinroth and S. Panichapong (Ed), Proceedings
    Second International Soil Classification Workshop.
    Part II. Soil Survey Division, Land Development
    Department, Bangkok, Thailand, 53-76.
Farmer, V.C., Russell, J.D. and Berrow, M.L. 1980.
    Imogolite and proto-imogolite allophane in spodic
    horizons: evidence for a mobile aluminium silicate
    complex in podzol formation. J. Soil Sci., 31,
    673-684.
Fedoroff, N. 1969. Caractères micromorphologiques des
    pédogenèses quaternaires en France. Etudes sur le
    Quaternaire dans le Monde. VIIIe Congrès Union
    Internationale pour l'Etude du Quaternaire, Paris,
    341-349.
Fedoroff, N. 1974. Classification of accumulations of
    translocated particles. In : G.K. Rutherford (Ed),
    Soil Microscopy, The Limestone Press, Kingston,
    Ontario, 695-713.
Frei, E. and Cline, M.G. 1949. Profile studies of normal
    soils of New York: II. Micromorphological studies
    of the grey-brown podzolic-brown podzolic soil
    sequence. Soil Sci., 68, 334-344.
Fridland, V.M. 1958. Podzolization and illimerization
    (clay migration). Soviet Soil Sci., 1, 24-32.
Gile, L.H. and Grossman, R.B. 1979. The desert soil
    monograph project. U.S.D.A.
Glinka, K.D. 1914. The great soil groups of the world
    and their development. (Transl. from German to
    English in 1927 by C.F. Marbut).
Gombeer, R. 1977. Potential and effective clay mobility
    in tropical soils. Clamatrops, Conference on
    classification and management of tropical soils.
    Kuala Lumpur, Malaysia, Malaysian Soc. Soil Sci.,
    1-11.
Hallsworth, E.G. and Crawford, D.V. (Ed), 1965.
    Experimental Pedology. Butterworths, London.
Hill, J.D. 1970. Quantitative micromorphological
    evidence of clay movement. In : D.A. Osmond and
    P. Bullock (Ed), Micromorphological Techniques
    and Applications. Soil Surv. Monogr. No.2,
    Harpenden, England, 34-42.

Isbell, R.F. 1980a. The argillic horizon concept and
    its application to the classification of tropical
    soils. In : Transactions Commissions IV and V,
    I.S.S.S., Kuala Lumpur, Malaysia, 1977. 127-134.
Isbell, R.F. 1980b. Genesis and classification of low
    activity clay Alfisols and Ultisols. In : B.K.G.
    Theng (Ed), Soils with Variable Charge.   New
    Zealand Soc. Soil Sci., 397-410.
Jamagne, M. 1972. Some micromorphological aspects of
    soils developed in loess deposits of Northern
    France. In : St. Kowalinski (Ed), Soil Micro-
    morphology, Warsaw, Poland, 559-582.
Khalifa, E.M. and Buol, S.W. 1968. Studies on clay skins
    in a Cecil (Typic Hapludalf) soil. I. Composition
    and genesis. Soil Sci. Soc. Am. Proc., 32, 857-861.
Kwaad, F.J.P.M. and Mücher, H.J. 1979. The formation
    and evolution of colluvium on arable land in
    northern Luxembourg. Geoderma, 22, 123-192.
McKeague, J.A., Guertin, R.K., Valentine, K.W.G.,
    Bélisle, J., Bourbeau, G.A., Michalyna, W., Hopkins,
    L., Howell, L., Pagé, F. and Bresson, L.M. 1980.
    Variability of estimates of illuvial clay in soils
    by micromorphology. Soil Sci., 129, 386-388.
McKeague, J.A., Wang, C., Ross, G.J., Acton, C.J., Smith,
    R.E., Anderson, D.W., Pettapiece, W.W. and Lord,
    T.M. 1981. Evaluation of criteria for argillic
    horizons (Bt) of soils in Canada. Geoderma, 25,
    63-74.
Mermut, A. and Jongerius, A. 1980. A micro-morphological
    analysis of regrouping phenomena in some Turkish
    soils. Geoderma, 24, 159-175.
Milfred, C.J., Hole, F.D. and Torrie, J.H. 1967. Sampling
    for pedographic modal analysis of an argillic
    horizon. Soil Sci. Soc. Am. Proc., 31, 244-247.
Miller, F.P., Wilding, L.P. and Holowaychuk, N. 1971.
    Canfield silt loam, a Fragiudalf: II. Micro-
    morphology, physical, and chemical properties. Soil
    Sci. Soc. Am. Proc., 35, 324-331.
Moorman, F.R. 1978. Report of Brazil meeting of the
    committee on the classification of Alfisols and
    Ultisols with low activity clays. In : M.N.
    Camargo and F.H. Beinroth (Ed), First International
    Soil Classification Workshop, Rio de Janeiro, Brazil,

45-63.

Nettleton, W.D., Flach, K.W. and Brasher, B.R. 1969.
    Argillic horizons without clay skins. Soil Sci.
    Soc. Am. Proc., 33, 121-125.

Nettleton, W.D., Witty, J.E., Nelson, R.E. and Hawley,
    J.W. 1975. Genesis of argillic horizons in soils
    of desert areas of the southwestern United States.
    Soil Sci. Soc. Am. Proc., 39, 919-926.

Parfenova, E.J., Mochalova, E.F. and Titova, N.A. 1964.
    Micromorphology and chemistry of humus-clay neo-
    formations in grey forest soils. In : A. Jongerius
    (Ed), Soil Micromorphology. Elsievier, Amsterdam,
    201-212.

Robinson, G.W. 1932. Soils, Their Origin, Constitution
    and Classification. Thomas Murby, London.

Rode, A.A. 1970. Podzol-forming process. (Translation
    of book published in Russian in 1937). Israel
    Program for Scientific Translations, Jerusalem.

Simonson, R.W. 1978. Soil survey and soil classification
    in the United States. Proceedings 8th Nat. Cong.
    Soil Sci. Soc. South Africa. Tech. Comm. 165,
    Dept. Agric. Tech. Services, 10-21.

Smeck, N.E., Wilding, L.P. and Holowaychuk, N. 1968.
    Genesis of argillic horizons in Celina and Morley
    soils of western Ohio. Soil Sci. Soc. Am. Proc.,
    32, 550-556.

Smeck, N.E., Ritchie, A., Wilding, L.P. and Drees, L.R.
    1981. Clay accumulation in sola of poorly drained
    soils of western Ohio. Soil Sci. Soc. Amer. J.,
    45.

Smith, G.D. 1963. Objectives and basic assumptions of
    the new soil classification system. Soil Sci., 96,
    6-16.

Smith, G.D., Sys, C. and van Wambeke, A. 1975. Applica-
    ion of Soil Taxonomy to soils of Zaire (Central
    Africa). Pédologie, 25, 5-24.

Smith, H. and Wilding, L.P. 1972. Genesis of argillic
    horizons in Ochraqualfs derived from fine textured
    till deposits of northwest Ohio and southeastern
    Michigan. Soil Sci. Soc. Am. Proc., 36, 808-815.

Soil Survey Staff, 1960. Soil Classification, 7th
    Approximation. Soil Conservation Service, U.S.D.A.,
    Washington, D.C.

Soil Survey Staff, 1975. Soil Taxonomy. Agriculture
    Handbook 436, U.S.D.A., Washington, D.C.,
Stoops, G. 1978. Some considerations on quantitative
    soil micromorphology. In : M. Delgado, (Ed),
    Micromorfologia de Suelos. Universidad de
    Granada, Spain, 1367-1384.
Torrent, J., Nettleton, W.D. and Borst, G. 1980. Genesis
    of a Typic Durixeralf of Southern California. Soil
    Sci. Soc. Am. J., 44, 575-582.
Wang, C. and McKeague, J.A. 1982. Illuviated clay in
    sandy Podzolic soils of New Brunswick. Can. J.
    Soil Sci., 62, (in press).
Wieder, M. and Yaalon, D.H. 1978. Grain cutans
    resulting from clay illuviation in calcareous
    soil material. In : M. Delgado (Ed), Micro-
    morfologia de Suelos. Vol.2. Universidad de
    Granada, Spain, 1133-1158.

PODZOLISATION AND THE SPODIC HORIZON

F. De Coninck[1] and D. Righi[2]

[1]Geologisch Instituut, Krijgslaan 271, B-900 Ghent, Belgium
[2]Laboratoire de Pédologie, Faculté des Sciences, 86022 Poitiers, France

ABSTRACT

In this review, podzolisation is considered as the different processes giving rise to the formation of soils with the macromorphological features of Spodosols, i.e. soils with a spodic or placic horizon.

The organic (O), organo-mineral (A1) and bleached (A2) horizons of all Spodosols have similar micro-features : plant remains, fungi, faecal pellets and pellets. A supplementary feature in A2 horizons is polymorphic organic matter with speckled aspect, which indicates a dissolution process of the organic matter.

Four kinds of B horizons can be distinguished: friable, cemented, nodular and placic. The predominant microfeatures of each are respectively : 1) the friable spodic horizons have mostly pellety material or units with polymorphic organic matter and fine mineral particles, 2) the cemented spodic horizons have organans or units with monomorphic organic matter, 3) in the nodular spodic horizons the Fe-rich parts have a dark reddish brown to reddish brown isotropic plasma, while the Fe-poor parts contain units of polymorphic or monomorphic organic matter, 4) the placic horizons have a reddish to dark reddish plasma; above this impervious horizon, a layer of plant remains and aggregates with pellets, small comminuted plant fragments and fungi, has developed.

From microprobe studies the following can be concluded : 1) in the friable spodic horizons Si, Al and possibly Fe are present inside the polymorphic units, but distributed in a different way; Si is part of the mineral grains only, whereas Al and Fe form parts of the mineral grains and are bound with the organic matter. 2) The monomorphic units have only Al, or Al and Fe, but no Si. 3) The placic horizon contains Fe and Al but the ratio Fe/Al is much higher than in monomorphic organic matter.

The formation of the friable spodic horizon is
due to two simultaneous processes : illuviation of
organo-Al or organo-Al+Fe complexes and biological
activity living on the remains of the numerous roots
and on the illuviating complexes.

The cemented spodic horizons have been formed
by accumulation of organo-metallic complexes that
constitute the cementing agent. The Fe-rich parts
of the nodular spodic horizons are the result of an
oxido-reduction process due to a high water-table
within the soil. The placic horizons are due to the
development of a reduced zone at the top of the soil,
as a result of either a cold and humid climate or the
presence of an impervious cemented spodic horizon.

INTRODUCTION

Studies of Spodosols and of podzolisation are
innumerable. These terms, however, can have various
meanings, and podzolisation can be defined in different
ways:

1. The process or processes giving rise to the formation
of a bleached horizon in the upper part of the mineral
soil. This concept does not necessarily include the
formation of a horizon of accumulation (Rode, 1970).

2. A process of weathering, which destroys the silicates
by complexing and acidifying organic substances, forming
organo-metallic compounds especially in association with
Al and Fe. The organo-metallic compounds may be immobile
and accumulate at the place where they have been formed,
or mobile and form a horizon after migration over a
certain distance (Pedro et al., 1969; Duchaufour,
1977).

3. A migration and accumulation of organic substance with
or without combination with Al and Fe.

4. The formation of a spodic horizon (Soil Survey Staff,
1975), i.e. a horizon with active organic substances,
bound with Al,or Al and Fe (organo-metallic compounds).
This spodic horizon can form either at the top of the
mineral soil or underneath a bleached (albic) horizon.
The concept does not, in principle, assume a weathering
of minerals, if Al or Al and Fe are available without

a process of weathering.
5. The formation of a placic horizon (Soil Survey Staff,
(1975) or iron pan.
        In this review, podzolisation is considered as the
different processes giving rise to the formation of
soils with the macromorphological features of Spodosols,
i.e. soils with a spodic or placic horizon. Their
morphological features are very variable.
        From the data compiled here, mechanisms are
proposed to explain these different morphological
features. To be valid, the mechanisms must be in
agreement with the structures, fabrics and pedological
features at all levels of observation. Throughout,
reference will be made to Spodosols but it is intended
that the discussion of mechanisms covers the formation
of B horizons of podzols that may not meet the requirements of the order of Spodosols in Soil Taxonomy (Soil
Survey Staff, 1975).

MORPHOLOGY OF SPODOSOLS
        The different horizons which may be present are :
organic horizons (O), organo-mineral horizons (A1),
bleached horizons (A2 or albic) and horizons of
accumulation (spodic and placic); their subdivisions
are based partly on Bullock and Clayden (1980).
        Several systems have been proposed to describe the
micromorphology of organic matter in soils. In this
review, the authors mostly make reference to the
system of De Coninck and al. (1974). This system has
the advantage of proposing only two terms, as follows:
The amorphous organic matter, i.e. the organic matter
in which the original organised structure cannot be
recognised microscopically, is subdivided into monomorphic and polymorphic forms. This subdivision is
intended to distinguish the amorphous organic matter
predominantly present in the form of pellets from the
amorphous organic matter forming continuous coatings
and bridges with a uniform appearance.

Organic Horizons

        (a) Non-hydromorphic Spodosols:
Non-hydromorphic Spodosols have been described as
having particular humus forms consisting of a number of

layers (Duchaufour, 1965, 1977). Under conifers or
heath, the humus form is mor or raw humus (Müller,
1887; Kubiena, 1953; Babel, 1975; Jongerius and
Rutherford, 1979). The formation of this humus form
is due to strong acidity and low biological activity,
and a direct relationship exists between this humus
form and the development of the Spodosol. The humus
form, moder, is characteristic of less well developed
Spodosols (Kubiena, 1953; Duchaufour, 1977).

However, field observations reveal that mor and
moder humus forms can be present in many other soils,
e.g. Inceptisols, Alfisols and Entisols (Soil Survey
Staff, 1975). On the other hand, in tropical and
subtropical regions, e.g. Malaysia (Malaysian Society
of Soil Science, 1977), Queensland, Australia (Brewer
and Thompson, 1977; Sys, pers. comm.), no organic
horizons are present, even in strongly developed
Spodosols.

Viewed microscopically, the same fundamental
features appear in mor and moder humus forms (De
Coninck et al., 1974; Babel, 1975) : fresh and
transformed plant remains, elements of fungi (hyphae,
sclerotia, spores), fresh and transformed faecal
pellets and other pellets of unknown origin (Fig. 1a
and b). Even in mor humus, most plant remains are
present as small angular fragments, mixed with faecal
pellets and pellets of other origin. This indicates
faunal activity. Some mineral particles are always
included with the organic material, often even within
plant fragments.

It can be concluded that there is no characteristic
humus form for non-hydromorphic Spodosols. Even when
there is a correlation between the humus form present
and profile development, care must be observed when
relating the present humus form to soil development.
Indeed, humus forms relate to a much shorter time
cycle than the soil profile type (Bullock and
Clayden, 1980).

(b) Hydromorphic Spodosols with peaty surface horizons:
Soil Survey Staff (1975) described three kinds of
organic matter under hydromorphic conditions. The

Fig. 1. (a,b) O1 horizon of Spodosol. Faecal
          pellets inside and outside plant remains,
          together with sclerotia. Many of the plant
          remains and faecal pellets are birefringent
          between crossed polarisers (b) and many
          mineral grains anisotropic; (c) A1 horizon
          of Spodosol. Fresh plant remains, discrete
          pellets, aggregates and sand grains. Partially
          crossed polarisers; (d) A2 horizon of
          Spodosol. Uncoated quartz grains, discrete
          pellets and aggregates. PPL; (e) A2 horizon
          of Spodosol. Uncoated quartz grains, pellets
          and aggregates of speckled polymorphic
          organic matter. PPL; (f) A2 horizon of
          Spodosol. Speckled polymorphic organic matter.
          Monomorphic coating (cracks are present)
          showing a punctuated morphology, suggesting
          that it is a stage of dissolution. Partially
          crossed polarisers.

Fig. 2. Friable B horizon of Spodosol: (a) Uncoated
          mineral grains and an aggregate composed of
          pellets, plant remains and hyphae. PPL;
          (b) Pellets containing birefringent patches
          (plant remains and mineral grains?). XPL;
          (c) Aggregates formed inside a root remain.
          PPL; (d) Well developed tubule. PPL; (e)
          SEM micrograph of polymorphic organic matter,
          plant remains and mineral particles; (f)
          SEM micrograph of tissue residues in polymorphic
          organic matter.

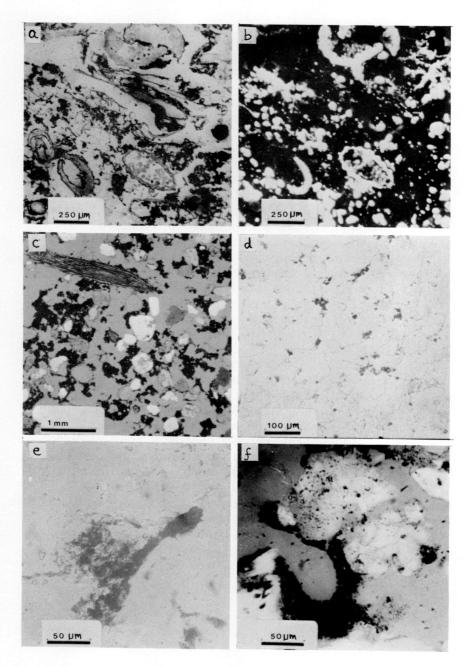

Fig. 1.

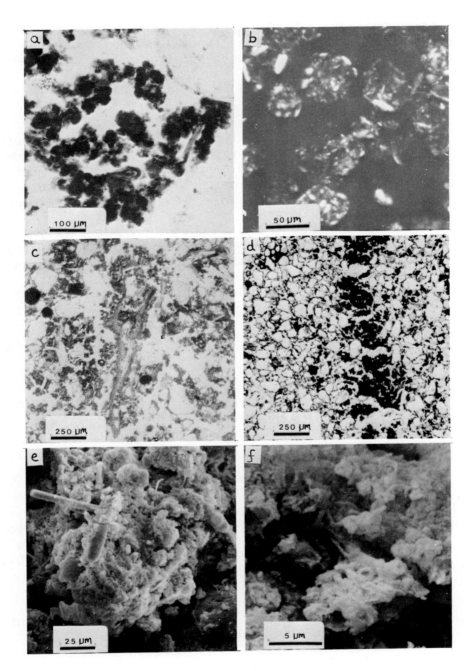

Fig. 2.

micromorphological features of these three forms of
peaty material do not differ fundamentally from those
of the organic matter in non-hydromorphic Spodosols.

Organo-mineral Horizons (A1) and Bleached Horizons (A2)

An A1 horizon normally has a dark greyish colour.
This colour is due to a mixture of discrete mineral grains,
especially quartz, and discrete organic particles. Under
the microscope, the organic particles appear to be mostly
plant remains and units of polymorphic organic matter
(De Coninck et al., 1974), as well as hyphae and sclerotia
(Fig. 1c).

In the A2 horizon, the amount of organic material is
usually small, with a composition similar to that in the
A1 horizon. A striking feature of the polymorphic organic
matter is its speckled aspect. Originally defined as
punctuated (Robin and De Coninck, 1978), it is composed of
black, very small (less than 5 μm) specks sometimes set
in a paler matrix. In some places, remains of monomorphic
coatings with the same speckled morphology are identifiable
(Fig. 1d, e and f).

In some A2 horizons, a duripan has been described
(Soil Survey Staff, 1975), probably due to cementation
with silica, but no information on the micromorphological
structure of this horizon is available.

Accumulation or B Horizons (Spodic and Placic Horizons)

(a) Macromorphology and physical characteristics:
In Spodosols, four different forms of horizons of
accumulation can be distinguished (De Coninck, 1980b):
1. Friable spodic horizon: - the horizon has a loose to
    friable consistence, a high porosity and many roots
    throughout. When the clay content exceeds about 10%,
    there may be a crumb structure.
2. Cemented spodic horizon: - the horizon has a con-
    tinuous ortstein with low porosity and few if any
    roots, at least in part of it; if roots are present,
    they form mats along the cracks.
3. Nodular spodic horizon: - the horizon shows an
    alternation, over short distances, of Fe-rich nodules
    or concretions and more friable Fe-poor material; the

Fig. 3.   Cemented B horizon of Spodosol: (a, b)
          Coatings of monomorphic organic matter with
          cracking pattern, around all mineral grains.
          PPL;   (c) SEM micrograph of thick monomorphic
          coating with smooth surface;   (d) SEM micro-
          graph of cracked monomorphic coating and
          hyphae (arrow) lying on it;   (e) SEM micrograph
          of monomorphic organic matter with irregular
          surface and absence of any layering;   (f)
          SEM micrograph of surface composed of small
          rounded bodies (not same profile as (e)).

Fig. 4.   (a) Cemented B horizon of Spodosol showing
          loose zones, continuous monomorphic coatings
          and aggregates;   (b) Placic horizon formed
          in cemented horizon with monomorphic coatings.
          Root mat above placic horizon contains plant
          remains and aggregates;   (c) Placic horizon
          formed in a horizon with monomorphic
          coatings. A very dark brown plasma with
          few cracks fills all the pores;   (d) Placic
          horizon formed in a loose (cambic?) horizon.
          The Fe accumulation is bright red, slightly
          birefringent and shows a flow pattern;   (e)
          Cemented B horizon of Spodosol. Monomorphic
          coating covering an older polymorphic
          aggregate;   (f) Sketch of (e):   1-quartz
          grain;   2-monomorphic organic matter;
          3-polymorphic organic matter;   4-pore.

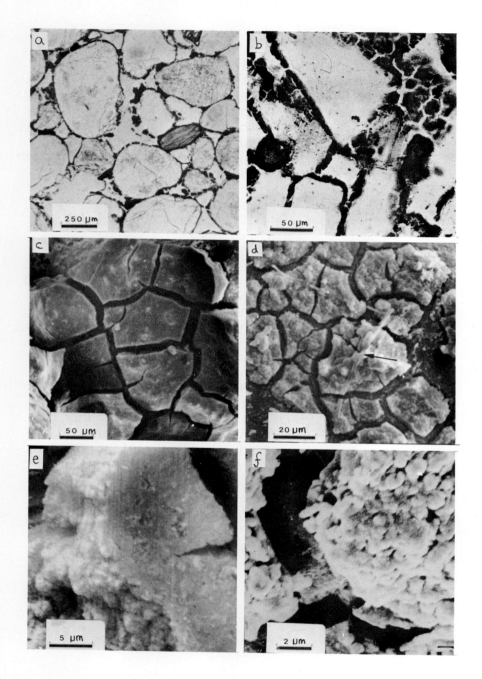

Fig. 3.

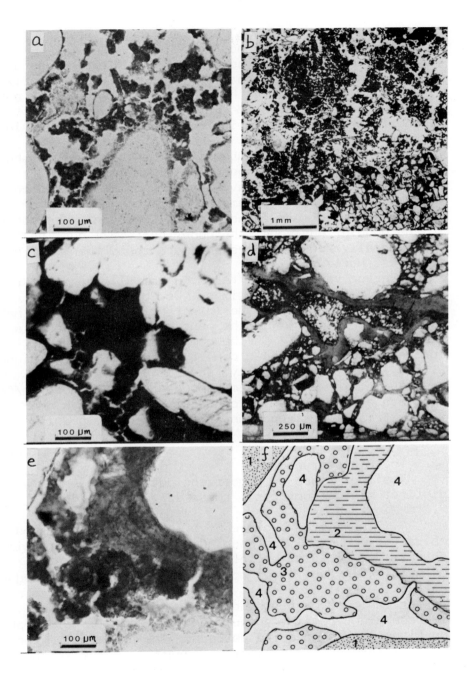

Fig. 4.

porosity is low and the roots are mostly limited to
the upper part of the horizon.
4. Placic horizon: - the horizon of accumulation has a
   placic horizon or is composed only of a placic horizon.
   A placic horizon is a thin, black to dark reddish pan
   cemented by Fe, Fe or Mn, or by Fe-organic matter
   complexes (Soil Survey Staff, 1975). It is impervious
   to roots.

In many Spodosols, subhorizons can be distinguished
in the friable, cemented and nodular horizons. The
friable horizons mostly have a darker upper B21 and paler
lower B22 horizon. The cemented horizons may have an upper
B21 horizon with more friable consistence and more roots.
In the nodular horizons, the upper part (B21 horizon)
has generally fewer nodules and more roots than the lower
part (B22 horizon).

Average values of different physical characteristics
vary widely for the subhorizons (De Coninck, 1980b) :
friable B horizons have the lowest bulk density and the
highest total porosity; the penetration resistance is
less than 30 kg/cm^2. In the cemented and nodular B
horizons the average bulk density is significantly higher,
with a lower total porosity, and the penetration
resistance exceeds 30 kg/cm^2 and can even reach 60 kg/cm^2;
on average, the hydraulic conductivity is much higher
in the friable horizon than in the cemented, nodular
and placic horizons.

(b) Microstructure:
The four types of B horizons outlined above have
distinct microstructures; these are discussed below.

1. The microstructure of friable spodic horizons is
characterised mainly by pedological features of organic
matter, described as pellets (Flach, 1960), pellety
material (Clayden, 1970) or units with polymorphic organic
matter (De Coninck et al., 1974). Polymorphic organic
matter can form coatings, discrete pellets or plasma
between plant remains and elements of fungi. Polymorphic
coatings lack a cracking pattern. The pellets are the
typical form of the polymorphic organic matter in the B
horizons. Their size varies from 20 to 200 µm; they
are brown to dark brown and opaque; their shape is mostly

rounded to ovoid with a rough surface (Fig. 2a and b).
The typical unit in the friable B horizons is the
aggregate, i.e. a porous cluster, which normally contains
pellets, opaque or birefringent plant fragments, hyphae
and sclerotia, and even faecal pellets. These aggregates
can also be found inside a hallow root fragment and be
part of aggrotubules or striotubules (Fig. 2c and d).
Mineral particles of clay and silt are mixed within the
pedological features containing organic matter in the
friable B horizons, wherever these particles are present
(Fig. 2b).
       Scanning electron (SEM) images show the heterogeneity
of the aggregates, with mineral (silt particles, phy-
toliths) and organic compounds (pellets, plant fragments,
hyphae, organic matter forming a plasma between the former
constituents) (Fig. 2e and f).

       2. The micromorphological characteristics of
cemented spodic horizons,called ortstein or alios, have
been described as chlamydomorphic fabric by Kubiena (1938),
granular with free grain organans by Brewer (1964, 1974),
and coated with monomorphic organic matter by De Coninck
et al., (1974). These coatings are opaque and yellowish
brown to very dark brown, depending on their thickness.
They have been described in ortstein or alios horizons in
different countries : De Bakker and Schelling (1966) in
The Netherlands, De Coninck and Righi (1969) and Robin
and De Coninck (1978) in France, FitzPatrick (1971) in
Scotland, Flach (1960) and Soil Survey Staff (1975) in the
U.S.A., and McKeague and Wang (1980) in Canada. They
cover mineral grains and pre-existing organic components
(pellets, plant fragments) and may form bridges at points
of contact between mineral grains. Under the polarising
microscope, the surface seems to be smooth (Fig. 3a and
b). SEM shows a strongly developed angular cracking
pattern at magnifications of about x200, separating
polygons with smooth surfaces (Fig. 3c). At about x600
magnification the polygons present an irregular surface
with hyphae on top (Fig. 3d). Most magnifications of
more than x2000 generally disclose a heterogeneous
stacking of rounded plates (Righi, 1977) (Fig. 3e) or
of rounded bodies (Robin, 1979) (Fig. 3f). The reasons
for these differences are not known at present. One of
the authors (Righi) is carrying out experiments 'in vitro'

to form monomorphic coatings.

Within the coatings, the composition appears uniform without any recognisable layering. This feature clearly distinguishes these coatings from argillans and ferri-argillans, which show a typical layering parallel to their surface (Van Ranst et al., 1980).

Where cemented horizons have parts, or a subhorizon, that are more friable and contain roots, units with polymorphic organic matter occur. These units have mostly replaced the monomorphic coatings, but they may develop in the pores between the coated skeleton grains (Fig. 4a).

3. In nodular spodic horizons the microfeatures of the Fe-rich and Fe-poor parts are distinctly different. The former have a dark reddish brown to reddish brown isotropic plasma, which covers the skeleton grains and may fill pores completely. The Fe-poor parts may have units with polymorphic or monomorphic organic matter.

4. In many placic horizons, two or more discrete pans may be distinguished, sometimes situated only 2 or 3 mm from each other (Fig. 4b). The microstructure of the placic horizon is characterised by an isotropic, reddish or dark reddish plasma in which cracks may be present. However, this cracking pattern is generally less strongly developed than in monomorphic coatings (Fig. 4c). This Fe-rich plasma may show a structure with a flow pattern (Fig. 4d).

Fig. 5. Microprobe analysis: (a) Cemented B horizon of Spodosol; (b) Friable B horizon of Spodosol. Image widths 200 μm. 1-quartz grain; 2-thick monomorphic coating; 3-aggregates of polymorphic organic matter.

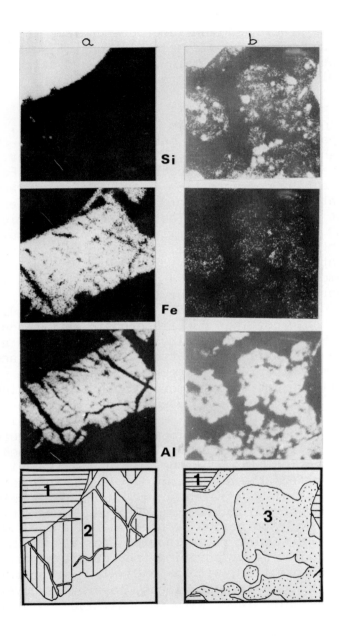

Fig. 5.

The placic horizons are more or less impervious to roots and root mats, and accumulations of aggregates with pellets, plant remains, hyphae and sclerotia commonly lie on their surfaces (Fig. 4b).

(c) Microprobe study:
In the friable spodic horizons (Fig. 5a), the units with polymorphic organic matter contain C, Al and Si throughout and possibly Fe. There is, however, a striking difference in the distribution of Si, Al and Fe. Si shows a high concentration in well defined places, indicating the presence of quartz and silicates; its amount is much lower in the polymorphic organic matter. Al and Fe are more uniformly distributed; they are present with Si in the silicates and have a uniform concentration in the polymorphic organic matter. In hydromorphic Spodosols, Fe is normally absent in the polymorphic organic matter (Righi, 1975).

In the cemented spodic horizons (Fig. 5b), the monomorphic coatings contain organic matter and Al, as well as Fe in freely drained conditions. The amounts of Si are too small to be detected clearly. Al and Fe are uniformly distributed throughout the coating. Fe is normally absent in hydromorphic conditions (Righi, 1975).

A placic horizon present in the middle of a spodic horizon (Placohumod) has been studied by Righi et al. (1982). The micromorphology, combined with quantitative determinations of Al and Fe, points to heterogeneity of the cementing plasma of the placic horizon (Fig. 6). Starting from the surface of the quartz grains, at least two layers with different composition can be distinguished: 1) an underlying brown organic layer, immediately adjacent to the skeleton grains, containing Fe and Al in a Fe_2O_3/Al_2O_3 molar ratio ranging from 0.4 to 2.5; 2) a more reddish layer covering the former, apparently poorer in organic matter, with a Fe_2O_3/Al_2O_3 molar ratio ranging from 5 to 50. In some places, layer 2 is covered with a black organic plasma with a Fe_2O_3/Al_2O_3 molar ratio ranging from 1 to 2.

This distribution of Fe, Al and organic matter is clearly different from that in monomorphic coatings.

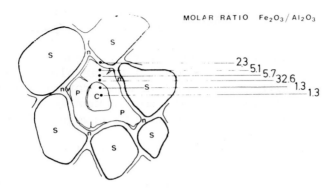

MOLAR RATIO Fe$_2$O$_3$/Al$_2$O$_3$

Fig. 6. Schematic drawing of a typical section
through a placic horizon (Placohumod,
Antwerp Campine, Belgium) with the value
of the molar ratio at different places.
s : skeleton grains; n: underlying layer;
p : intermediate reddish layer; c : core
with black organic plasma; v : void.

ORIGIN OF THE DIFFERENT SPODIC HORIZONS
Friable Spodic Horizons
    Different theories for the origin of the units with
polymorphic organic matter have been proposed.
(a)   Biological activity.
(b)   Physico-chemical formation by adsorption of organic
      compounds on Fe and Al hydroxides.
(c)   Flaking of cracked illuvial coats off mineral grains.

    (a) Biological activity:
    This process can form pellety material in two ways :
a fraction of plant remains, ingested by fauna, forms
faecal pellets and pellets by transformation of the faecal
pellets; the part not taken up, but comminuted into
small pieces, is transformed into dark pellets (De Coninck
and Laruelle, 1964; De Bakker and Schelling, 1966;
De Coninck and Righi, 1969; Eswaran et al., 1972; Conry
et al., 1972; De Coninck et al., 1974; Robin and
De Coninck, 1978).
    Indications of the presence of animals in the friable
B horizons are numerous.
1. Friable B horizons always contain roots. There is thus

a continuous decay of roots; if dead roots were transformed only biochemically or physico-chemically, large
fragments of tissue should be recognisable. In reality
however, such fragments are very rare, whereas the organic
units generally contain very small, angular plant remains
intimately mixed with polymorphic organic matter and the
fine mineral fraction : an active microfauna is the most
effective agent to produce such a microstructure (Babel,
1975).
2. Burrowing animals (Enchytraeidae, small earthworms)
form characteristic aggrotubules or striotubules.
The aggregates within are identical to those outside the
tubules.
3. Determination of mean residence times (MRT) with ^{14}C.
      Guillet and Robin (1972), Righi and Guillet (1976) and
De Coninck (1980a) have determined the MRT in friable and
cemented horizons in several profiles from France and
Belgium. Their data show clearly that there is a direct
relationship between consistence, micromorphological
features and the MRT. The organic matter in horizons
with loose consistence and polymorphic units is much
younger than in horizons with cemented consistence and
monomorphic units. The differences in age indicate that
the turn-over in polymorphic organic matter is much faster
than in monomorphic organic matter. These differences in
MRT between polymorphic and monomorphic organic matter
suggest that the nature of these two forms of organic
matter is different.
4. Righi (1981) has studied the distribution of the
different fractions of organic matter (fulvic acids,
humic acids, non-extractable organic matter) in aggregates
of different size, originating from horizons either with or
without pellety structure. The distribution of the non-
extractable organic matter is identical in aggregates
from A1 and spodic horizons. This suggests that the same
mechanism is responsible for the formation of all
aggregates, i.e. faunaturbation. On the other hand, the
extractable organic matter has the same composition in
both friable and cemented B horizons. This suggests
that two processes produce the pellety microstructure
of the friable B horizons : the microstructure itself
is the result of the action of the fauna but the aggregates contain a large amount of illuvial material,

associated with the fine mineral fraction and root
remains.

(b) Physico-chemical formation by adsorption of
organic compounds on Fe and Al hydroxides:
Boudot and Bruckert (1978) and Bruckert and Selino
1978) have attempted to explain the origin of the pellets
through a study of their composition. They consider that
a direct relationship exists between the pellety microstructure and the amount of organo-ferric complexes. The
mechanism of formation is explained as follows : soluble
organic compounds are adsorbed at the surface of the clay
particles and amorphous metallic hydroxides. This
adsorption modifies the physico-chemical properties of
the hydroxides, which acquire the characteristic pellety
microstructure. Bruckert and Selino (1978) however observe
a "remarkable similarity in shape, and even in composition,
between the microaggregates of biological origin at one
hand, and the microaggregates proceeding from the assumed
physico-chemical process at the other hand".
Notwithstanding the arguments in favour of a biological origin and the absence of any clear explanation
of the physico-chemical process, the authors use the
composition of the aggregates as an argument to reject
the biological origin and to accept the physico-chemical
origin.
This theory fails to explain the presence of the large
amount of small discrete plant remains in the B horizons
and the fate of the roots present. It also does not
explain the low MRT of the organic carbon.

(c) Flaking of cracked illuvial coats off mineral
grains:
Flach (1960) and Franzmeier and Whiteside (1963)
propose that the aggregates are formed by the flaking
off of cracked illuvial coats from mineral grains. This
concept assumes that aggregates should be observed only
in horizons with strongly developed monomorphic coatings.
This does not agree with general observation in many
profiles. Moreover, as in the case of the physico-chemical
theory, it does not explain the presence of the plant
remains and the low MRT. However, it may be part of the

origin of some features (see below).

Cemented Spodic Horizons

The morphology of monomorphic coatings suggests deposition from an aqueous phase, covering all pre-existing components with a continuous layer and smoothing out the irregularities of their surfaces. The rounded bodies or plates which build up the coating and the very large volume occupied by the cracks indicate that these coatings are formed by immobilisation of compounds with high water content; this means that at the moment of their deposition they must be in a gel stage. The release of the water causes strong shrinking with formation of the cracking pattern.

The coatings are built up by successive additions probably over a long period (De Coninck, 1980a). If a complete and irreversible desiccation occurs after each deposition, the successive additions should presumably fill the cracks formed in the preceding deposit. It appears, however, that the coating material retains the property to take up water again, at least to some extent, integrating new additions into the material already in place, Thus the monomorphic coatings appear to remain in a state capable of water uptake to some extent, but not to the point at which they become mobile again.

The presence of Al or Al and Fe, without Si, in these coatings, as shown by the microprobe study, and the chemical analysis of these B horizons (Righi and De Coninck, 1977; De Coninck, 1980 a , c) indicate that the monomorphic coatings are composed of organo-Al or organo-Al + Fe complexes. They seem to constitute the fraction that has migrated out of the topsoil and accumulated in the B horizons.

Nodular Spodic Horizons

The iron nodules of these B horizons have large amounts of free Fe (De Coninck, 1980c) and relatively low organic matter contents, so that an accumulation of Fe as an organo-metallic complex is impossible. Indeed, the Fe/C molar ratio considerably exceeds the highest possible ratio over which these complexes are mobile (De Coninck, 1980a). Moreover, field study shows that the profiles have developed in drainage conditions in which the water-table

stands periodically at the level of the nodular horizon.
The same mottles and nodules can also be found in soils
without any accumulation of organic matter. It is
evident that the fluctuating water-table causes the
formation of a horizon with hardened Fe-rich parts
(nodules and mottles) and loose or friable Fe-poor
parts (bleached). This stage in formation precedes
podzolisation, but when organo-metallic compounds
accumulate in the loose Fe-poor parts of the horizon,
the hardened Fe-rich features are incorporated into
the B horizon where the presence of the organic matter may
enhance their hardening.

FORMATION OF SPODOSOLS WITH FRIABLE AND CEMENTED HORIZONS
Formation of Active Organo-metallic Compounds

Active organic substances are substances with a
high pH-dependent charge, a large surface, and high
water retention (Soil Survey Staff, 1975). Since in
many Spodosols these active organic substances
accumulate at some depth, it follows that they are mobile
under some conditions but may lose mobility under others.
They can complex Al and Fe, and the organo-metallic
complexes formed remain mobile up to a certain metal/
organic matter ratio. Above this ratio the complexes
become immobile (McKeague et al., 1981). The complexed
Al and Fe can either be present in available form in
the parent material, or supplied by mineral weathering
(De Coninck, 1980a).

The shape of monomorphic coatings and their chemical
composition, as shown by microprobe studies and chemical
analyses (Righi, 1977; Righi and De Coninck, 1977;
De Coninck, 1980c) indicate that they consist of
organo-metallic complexes with high water retention, which
have accumulated following migration.

The features of the polymorphic units, on the other
hand, cannot be explained solely by a process of
illuviation of organo-metallic compounds. Their shape and
composition can be explained only through intervention
of fauna, living on plant remains and illuviated organo-
metallic compounds.

Stages of Development

In the beginning of the evolution of most parent
materials the amounts of available Al and Fe are
sufficient to prevent illuviation of mobile organic
substances, but a horizon with some organo-metallic
complexes is formed at the top of the mineral soil.
Once at least some organo-metallic complexes are mobile,
due either to insufficient amounts of Al and Fe or to
the production of a larger amount of mobile substances,
an A2 horizon is gradually formed by removal of Al and
Fe. Where the mobilised organo-metallic complexes are
immobilised again, a B horizon develops. The speckled
aspect of the organic matter and the presence of remains
of monomorphic coatings with the same speckled composition in the A2 horizon suggests a dissolution process
(Robin and De Coninck, 1978) and thus a gradual deepening
of the A2 horizon.

Evolution is determined by the relative intensities
of two opposing processes : the continuing illuviation
of mobile organic compounds and pedoturbation. The
first tends to cement the horizon, the second to keep it
friable.

If pedoturbation keeps pace with illuviation of
organo-metallic compounds, the horizon retains a loose
consistence, allowing regular root development. If
illuviation is greater than the mixing activity of the
fauna, the B horizon is gradually cemented and, after
some time, the progressive cementation impedes the
penetration and development of roots, and consequently
also the biological activity. This results in the
fossilisation of the horizon. Thus, soils with a
cemented B horizon passed through a stage with polymorphic organic matter followed by a stage with monomorphic
organic matter. The succession of the two stages is
evident in many profiles, in which monomorphic
coatings cover polymorphic units (Fig. 4e).

In many Spodosols a fourth stage in the evolution
may be recognised. Roots develop preferentially above
and in the upper part of a cemented horizon. This
accumulation of roots favours biological activity with
formation of polymorphic units from the roots and the
pre-existing monomorphic organic matter. Micromorphological evidence of the uptake of illuviated compounds

lies in the disappearance of monomorphic units where
roots develop in a cemented horizon. The uptake of this
monomorphic material may be facilitated by the flaking
of coatings from the skeleton grains (Flach, 1960).

Ecology of the Types of Spodic Horizons
        The presence of either a friable or a cemented
spodic horizon is not accidental. The appearance of
one, exclusive of the other, is related to the nature
of the environment. Righi (1969) and Righi and De Coninck
(1974) showed that with more impeded drainage a cemented
spodic horizon may grade into a friable one. In freely
drained conditions, Robin (1979) observed cemented spodic
horizons in very poor quartzitic sands, but with an
increasing amount of weatherable minerals, especially
trioctahedral chlorite, the spodic horizon gradually
becomes more friable with pellety microstructure. This
change in composition of the parent material and in soil
development is accompanied by a change in vegetation
(broad-leaf forest on friable Spodosols, conifers on
cemented Spodosols).
        Guillet (1972),studying soils in the Vosges,noted
that replacement of conifers by heather with Calluna
through anthropic influence accentuates the process of
podzolisation; the spodic horizons had higher organic
matter content and were more cemented.
        It can be concluded that all factors limiting the
production, persistence and mobility of organic substances
tend to slow down the formation of a cemented horizon. On
the other hand, factors limiting root development and
faunal activity favour cementation of the spodic horizons.

FORMATION OF SOILS WITH A PLACIC HORIZON
        Placic horizons (Soil Survey Staff, 1975) are
fairly common in regions with humid temperate to cold
climate. They are characterised by very high Fe, Fe
and Mn, or Fe and organic matter contents. Therefore,
the conditions of the environment allow a strong specific
accumulation of Fe. Placic horizons have formed at the
top, within or beneath a spodic horizon (e.g. Placaquods,
Placohumods, Placorthods), in the mineral soil beneath
deep peat deposits (Histosols), in the C horizon of
mineral soils (e.g. Placandepts, Placaquents) (Crampton,

1963; McKeague et al., 1967 and 1968; Soil Survey
Staff, 1975; Guillet et al., 1976).

Some detailed studies on the composition and genesis
of placic horizons are given in literature, three of which
are discussed below in some detail.

McKeague et al. (1967), investigating an iron pan
humic Podzol from Newfoundland, conclude that Fe is present
as Fe-fulvic acid complexes. However, this conclusion is
not supported by their chemical data, since the amount of
organic matter (15 to 17%) is evidently too small to
completely complex the amount of Fe (13.5%). Therefore,
most Fe is not in the form of Fe-fulvic acid complexes,
but rather as free Fe oxides or hydroxides.

From a study of other soils with placic horizon,
McKeague et al., (1968) conclude that Fe and Mn have
probably been transported as divalent ions, oxidised and
precipitated.

In the placic horizon of the Placohumod from the
Antwerp Campine (Righi et al., 1982), the content of Fe_2O_3
exceeds several times the amount of Fe which may be
transported by the organic matter present (Schnitzer and
Kahn, 1972; Gamble and Schnitzer, 1973). The hypothesis
of the authors is that the cemented horizon creates an
impervious zone, causing an accumulation of roots and a
perched water-table. The combination of organic matter and
water saturation brings about an environment which is
reducing for Fe^{+3}, bound to the organic matter, solubi-
lising the iron as Fe^{+2}. This Fe^{+2} is transported downwards but is oxidised again where the environmental
conditions become more oxidising.

In the three studies noted above the amount of Fe is
evidently too high to explain its migration as an organo-
metallic complex; the only logical explanation is therefore that Fe is translocated in reduced form and that
the formation of the placic horizon is due to a reprecipitation of Fe after reoxidation.

This concept agrees with that of several other authors
(Crompton, 1956; Crampton, 1963; FitzPatrick, 1971;
Duchaufour, 1976) who consider that a placic horizon forms
in a zone of strong changes in redox potential at the
boundary between the H_2O-saturated upper and the aerated
lower horizons.

In conclusion, the authors consider that two kinds of
placic horizons can be distinguished : 1. placic

horizons whose differentiation is unrelated to a process of
podzolisation (e.g. placic horizons of Histosols,
Placaquepts and some Aquods); 2. placic horizons whose
formation is genetically related to the development of
the spodic horizon (Placohumods, Placorthods).

In both cases the conditions of the environment bring
about a reduction of insoluble Fe^{+3} to soluble Fe^{+2}. In
the first case, these reducing conditions are caused by
the external climatic conditions, high rainfall and
low temperature giving a reducing environment at the top
of the soil. In the second case, the reducing conditions
are the result of the development of the spodic horizon
which forms an impervious horizon within the soil (soil
with monomorphic units).

REFERENCES
Babel, U. 1975. Micromorphology of soil organic matter.
    In: J. E. Gieseking (Ed), Soil Components. Vol. 1.
    Organic Components. Springer-Verlag, New York,
    369-473.
Boudot, J.P. and Bruckert, S. 1978. Complexes organo-
    metalliques et structures microagregées des sols
    chloriteux du système schistograuwackeux Vosgien.
    Sci. du Sol, 1, 31-40.
Brewer, R. 1964. Fabric and Mineral Analysis of Soils.
    John Wiley and Sons, New York, 470 pp.
Brewer, R. 1974. Some considerations concerning micro-
    morphological terminology. In: G.K. Rutherford
    (Ed), Soil Microscopy. The Limestone Press,
    Kingston, Ontario, 28-48.
Brewer, R. and Thompson, C.H. 1977. Two subtropical
    podzols. Clamatrops. Abstracts of Papers.
    Conference on classification and management of
    tropical soils. Kuala Lumpur, Malaysia, 12 pp.
Bruckert, S. and Selino, D. 1978. Mise en évidence de
    l'origine biologique ou chimique des structures
    microagregées foisonnantes des sols bruns ocreux.
    Pédologie, 28, 46-59.
Bullock, P. and Clayden, B. 1980. Morphological properties
    of Spodosols. In: B.K.G. Theng (Ed), Soils with
    Variable Charge. Soil Bureau, D.S.I.R.O., Lower
    Hutt, N.Z., 45-65.
Clayden, B. 1970. The micromorphology of ochreous B
    horizons of sesquioxidic brown earths developed in

upland Britain.   In: D.A. Osmond and P. Bullock
(Ed), Micromorphological Techniques and Applications.
Soil Survey Tech. Monogr. 2, Harpenden, England,
53-67.

Conry, M.J., De Coninck, F., Bouma, J., Cammaerts, C. and
Diamonds, J.J. 1972. Some brown podzolic soils in
the west and south-west of Ireland. Proc. Royal
Irish Academy, 27, 13, No.21, 359-401.

Crampton, C.B. 1963. The development and morphology of
ironpan podzols in mid and south Wales. J. Soil
Sci., 14, 282-302.

Crompton, E. 1956. The environmental and pedological
relationships of peaty gleyed Podzols. Trans. 6th
Int. Congr. Soil Sci. E, 155-161.

De Bakker, H. and Schelling, J. 1966. Systeem van bodem-
classificatie voor Nederland. De hogere niveaus.
Neth. Soil Surv. Inst. Wageningen.

De Coninck, F. 1980a. Major mechanisms in formation of
spodic horizons. Geoderma, 24, 101-128.

De Coninck, F. 1980b. The physical properties of Spodosols.
In: B.K.G. Theng (Ed), Soils with Variable Charge.
Soil Bureau, D.S.I.R.O., Lower Hutt, N.Z., 325-349.

De Coninck, F. 1980c. Genese en eigenschappen van Podzolen.
Aggregaat proefschrift. Rijksuniversiteit Gent,
Belgium.

De Coninck, F. and Laruelle, J. 1964. Soil development in
sandy materials of the Belgian Campine. In: A.
Jongerius (Ed), Soil Micromorphology. Elsevier,
Amsterdam, 169-188.

De Coninck, F. and Righi, D. 1969. Aspects micromorphologiques de la podzolisation en forêt de Rambouillet.
Sci. du Sol, 2, 57-77.

De Coninck, F., Righi, D., Maucorps, J. and Robin, A.M.
1974. Origin and micromorphological nomenclature of
organic matter in sandy Spodosols. In: G.K. Rutherford (Ed), Soil Microscopy. The Limestone Press.
Kingston, Ontario, 263-280.

Duchaufour, P. 1965. Précis de Pédologie. 2nd Edition.
Masson, Paris.

Duchaufour, P. 1976. Atlas Ecologique des Sols du Monde.
Masson, Paris.

Duchaufour, P. 1977. Pédologie, 1 : Pédogénèse et
Classification. Masson, Paris.

Eswaran, H., De Coninck, F. and Conry, M.J. 1972. A
    comparative micromorphological study of light and
    medium textured podzols. In: St.Kowalinski (Ed), Soil
    Micromorphology. Warsaw, Poland, 269-285.
FitzPatrick, E.A. 1971. Pedology, a Systematic Approach to
    Soil Science. Oliver and Boyd, Edinburgh.
Flach, K.W. 1960. Sols bruns acides in the north eastern
    United States; genesis, morphology and relationships
    to associated soils. Ph.D. Thesis, Cornell Univ.,
    Ithaca, New York.
Franzmeier, D.P. and Whiteside, E.P. 1963. A chrono-
    sequence of podzols in northern Michigan. Michigan
    Agric. Exp. Sta., Quarterly Bull., 46, 2-57.
Gamble, D.S. and Schnitzer, M. 1973. The chemistry of
    fulvic acid and its reactions with metal ions.
    In: P.C. Singer (Ed), Trace Metals and Metal-Organic
    Interactions in Natural Waters. Ann Arbor Science Publ.
Guillet, B. 1972. Relation entre l'histoire de la
    végétation et la podzolisation dans les Vosges
    Thèse, Univ. de Nancy.
Guillet, B. and Robin, A.M. 1972. Interpretation de
    datations par le ^{14}C d'horizons Bh de deux podzols
    humo-ferrugineux, l'un formé sous callune, l'autre
    sous chênaie-hêtraie. C.R. Acad. Sci., Série D,
    274, 2859-2862.
Guillet, B., Rouiller, J. and Souchier, B. 1976. Accumula-
    tions de fer (lépidocrocite) superposées à des
    encroûtements ferrimanganiques dans des sols hydro-
    morphes vosgiens. Bull. Soc. Geol. France, 18,
    55-58.
Jongerius, A. and Rutherford, G.K. (Ed), 1979. Glossary
    of Soil Micromorphology. Pudoc, Wageningen, The
    Netherlands.
Kubiena, W.L. 1938. Micropedology. Collegiate Press Inc.
    Ames, Iowa.
Kubiena, W.L. 1953. The Soils of Europe. Thomas Murby
    and Co., London.
Malaysian Society of Soil Science. 1977. Characteristics
    of some soils in Sabah and Sarawak. Tour guide.
    Clamatrops Conference, Kuala Lumpur, Malaysia.
McKeague, J.A., Schnitzer, M. and Heringa, P.K. 1967.
    Properties of an ironpan humic podzol from
    Newfoundland. Can. J. Soil Sci., 67, 23-32.

McKeague, J.A., Damman, A.W.R. and Heringa, P.K. 1968.
    Iron-manganese and other pans in some soils of
    Newfoundland. Can. J. Soil Sci., 48, 243-253.
McKeague, J.A. and Wang, C. 1980. Micromorphology and
    energy dispersive analysis of ortstein horizons of
    podzolic soils from New Brunswick and Nova Scotia,
    Canada. Can. J. Soil Sci., 60, 9-21.
McKeague, J.A., Franzmeier, D.P. and De Coninck, F. 1981.
    Spodosols. In: L.P. Wilding, N.E. Smeck and G.F.
    Hall (Ed), Soil Genesis and Morphology. Elsevier
    (in press).
Müller, P. 1887. Studien über die Natürlichen Humusformen.
    Springer-Verlag, Berlin.
Pedro, G., Jamagne, M. and Begon, J.C. 1969. Mineral
    interactions and transformations in relation to
    pedogenesis during the Quaternary. Soil Sci., 107,
    462-469.
Righi, D. 1969. Aspects morphologiques et physico-chimiques
    de la pédogénèse en forêt de Rambouillet (France).
    Thèse de spécialité, Faculté des Sciences, Orsay,
    Paris.
Righi, D. 1975. Etude au microscope électronique à
    balayage de champ et au micro-analyseur à sonde
    électronique des revêtements et des agrégats organi-
    ques d'horizons B spodiques. Sci. de Sol Bull.
    A.F.E.S., 4, 315-321.
Righi, D. 1977. Génèse et évolution des podzols et des
    sols hydromorphes des Landes du Médoc. Thèse,
    Doctorat es Sciences, Univ. de Poitiers, France.
Righi, D. 1981. Relations entre l'illuviation de matière
    organique, l'activité de la microfaune et les
    structures d'horizons B de sols podzolisés du
    plateau de Millevaches (France). Pédologie, (in
    press).
Righi, D. and De Coninck, F. 1974. Micromorphological
    aspects of Humods and Haplaquods of the Landes du
    Médoc, France. In: G.K. Rutherford (Ed). Soil
    Microscopy. The Limestone Press. Kingston, Ontario,
    567-588.
Righi, D. and Guillet, B. 1976. Datation par le ^{14}C
    naturel de la matière organique d'horizons spodiques
    de podzol des Landes du Médoc (France). Soil organic
    matter studies. Int. Atomic Energy Agency, Vienna.
    II, 187-192.

Righi, D. and De Coninck, F. 1977. Mineralogical
    evolution in hydromorphic sandy soils and podzols
    in "Landes du Médoc", France. Geoderma, 19,
    339-359.
Righi, D., Van Ranst, E., De Coninck, F. and Guillet, B.
    1982. Microprobe study of a Placohumod in the
    Antwerp Campine (Belgium). Pédologie, (in press).
Robin, A.M. 1979. Génèse et évolution des sols podsolisés
    sur affleurements sableux du Bassin Parisien.
    Thèse, Doctorat es Sciences, Univ. de Nancy I,
    France.
Robin, A.M. and De Coninck, F. 1978. Micromorphological
    aspects of some podzols in the Paris Basin (France).
    In: M. Delgado (Ed). Micromorfologia de Suelos.
    Universidad de Granada, 1019-1050.
Rode, A.A. 1970. Podzol-forming process. Acad. Sci.
    U.R.S.S., Dokuchaev Inst. Transl. from Russian
    by A. Gourevitch. Israel Program for Scientific
    Translations, Jerusalem.
Schnitzer, M. and Kahn, S.U. 1972. Humic Substances
    in the Environment. Marcel Dekker, New York.
Soil Survey Staff, 1975. Soil Taxonomy. Agriculture
    Handbook No.436, U.S.D.A. Soil Conservation Service,
    Washington, D.C.
Van Ranst, E., Righi, D., De Coninck, F., Robin, A.M. and
    Jamagne, M. 1980. Morphology, composition and
    genesis of argillans in soils. J. Microscopy, 120,
    353-361.

MICROMORPHOLOGY OF THE OXIC HORIZON

G. Stoops

Laboratorium voor Mineralogie, Petrografie en
Micropedologie,Rijksuniversiteit Gent, Krijgslaan 271,
B-9000 Gent, Belgie

ABSTRACT

Although the oxic horizon is mainly defined
physico-chemically, several typical micromorphological
characteristics can be identified. As a result of the
pedoplasmation process the soil material is remarkably
homogeneous. In most well developed oxic horizons a
microped-structure can be observed, which can be formed
in several ways and is responsible for the fluffy
consistency of Oxisols in the field.
Virtually no silt-size mineral grains are present,
and the sand fraction contains no weatherable minerals.
As a result of its typical mineralogy (mixture of
kaolinitic clay and Fe and Al oxihydrates) the
birefringence fabric of the fine material is undifferentiated (isotropic) or weakly scaly. In some
cases a weak circular fibrous or a crescentic fibrous
fabric may be noted. Features related to clay illuviation
are virtually absent. In many Oxisols allochthonous or
disorthic lateritic nodules occur. They show a large
range of different internal fabrics, which can be related
to their mode of formation. An in situ formation of
gibbsitic features is also occasionally observed in oxic
horizons.
The micromorphological criteria for determining
oxic soil material are discussed, and the role of micropedology in the study of this material is emphasized.

INTRODUCTION
The oxic horizon is the diagnostic horizon of the
Oxisols (Soil Survey Staff, 1975). It is defined mainly on
physico-chemical characteristics. In contrast to other
diagnostic horizons the oxic horizon is not (or only to
a limited extent) determined by the expression of current
pedogenetic processes, but rather by the nature of the

material; mainly by its high degree of weathering, which in turn depends on several factors such as the type of parent material, the intensity of the weathering process and the geomorphological position.

The most important properties of oxic horizons are the lack of weatherable minerals in the sand and silt fractions and a clay fraction consisting of a mixture of kaolinite and oxihydrates of Fe and Al. These specific properties determine the physicochemical and morpho-logical characteristics of oxic horizons. In addition, by definition, an oxic horizon has a clay content of more than 15%. Soils with less clay can be classified only as oxic subgroups (e.g. Orthoxic Quartzipsamments).

Oxisols are very widespread in the humid tropical zones of Africa, Asia and Central and South America. As population increases, an understanding of the genesis and behaviour of these soils will be increasingly important, with a view to their future management. Furthermore, a better insight into their genesis is important in order to extend our knowledge of the geomorphological evolution of these areas.

Oxisols can be correlated partly with latosols and lateritic soils, the ferralsols of the INEAC and FAO classifications and the ferralitic soils of the French and Soviet classifications. Oxic soil materials have been described since the beginning of this century (Harrassowitz,(1930); Vageler,(1930); Harrison,(1933)). the first micromorphological study (except for some occasional observations) was only published in 1948 by Kubiena.

Most of the micromorphological features discussed in this paper may be observed also in oxic materials which for one reason or another do not meet the definition of a true oxic horizon (e.g. in oxic subgroups).

FORMATION AND HOMOGENISATION OF THE SOIL MATERIAL

One of the most important characteristics of well drained Oxisols (and most oxic subgroups) is their homogeneity. This homogeneity, which is present even in the deepest soil horizons, is caused mainly by mechanical processes such as mass movement, animal activity and rooting.

The heterogeneous saprolite, formed mainly by chemical weathering of the parent rock, is transformed into a homogeneous soil material, with a similar mineralogical and chemical composition but a different texture. Micromorphologically this is expressed by a loss of the original rock structure and a slow breakdown of the pseudomorphs after primary minerals, except those that are too hard and coherent (e.g. well crystallised goethite pseudomorphs). Kaolinite booklets, for instance, are transformed easily to a mass of clay-size particles stained by absorbed Fe oxihydrates.

In granitic rocks the transition is gradual. Investigations of soils from several regions have shown that mechanical destruction of the original rock structure may be partly caused by exfoliation of weathering biotites, followed by a collapse of pores formed by congruent dissolution of some components (e.g. amphiboles). On basalts, serpentinites, marbles etc. the transition is generally more abrupt and the chemical weathering front is followed immediately by mechanical disintegration. In addition, illuviation of clay from the overlying soil is common.

This process was called pedoplasmation by Flach et al. (1968). It is especially a feature on saprolites with high chroma differences (Folster, 1971). Romashkevich (1964, 1974) described it as the transformation of the weathering crust into the active zone of soil formation. More recently Beaudou and Chatelin (1979) pointed out that pedoplasmation starts in preferential areas such as ancient diaclases, veins and heterogeneities, and that it may be accompanied by illuviation of several soil components. In their view a gap exists between the alteration and the pedoplasmation front, both not necessarily moving at the same speed.

Homogenisation of the soil profile constantly takes place by activity of the soil mesofauna and the growth of roots. Termites seem to be particularly important in this process. This strong tendency towards homogenisation is opposed to the other pedogenic processes (e.g. illuviation) which tend to develop a vertical anisotropy in the soil.

MICROMORPHOLOGICAL ASPECTS OF THE OXIC HORIZON

Microstructure
        In the field, most oxic materials are massive or
have only a weakly developed coarse blocky structure. The
material disintegrates however to very fine, stable
granules under mild finger pressure. This granulation
is also responsible for the characteristic fluffy
consistence of most Oxisols. In thin section these
granules have been recognized for a long time in
ferrallitic soils (e.g. Fölster, 1964; Condado, 1969;
Bennema et al., 1970; Verheye and Stoops, 1975, and later
by many ORSTOM pedologists), and the characteristic
microped structure (Fig. la and b) has been considered as
one of the typical features of many Oxisols (Buol and
Eswaran, 1978). The last authors refer to it as an
agglutinic special related distribution pattern (Eswaran
and Baños, 1976). They also state that the expression
of the agglutinic SRDP is a function of the degree of
weathering (best expressed in the acric group) and the
climate (better expressed in the ustic than in the
udic group). Although these features have been recognised for a long period, only during the last decade
have more detailed investigations been made on the real
expression of this phenomenon and its genesis. ORSTOM
pedologists, in particular, have contributed to
investigations into its mode of formation (Beaudou, 1972;
Pedro et al., 1976; Chauvel, 1977; Muller, 1977).
        Muller (1977) in a study of red ferrallitic soils of
the central Cameroons distinguishes five genetically
different types of micropeds: network, zoogenetic,
complex, relict and ferritic. The last two, however,
can also occur embedded in a matrix, and therefore must
be considered also as micronodules.
        The formation of network micropeds (micropeds de
réseau), as described by Muller (1977), can be summarised
as follows. In deeper horizons birefringent streaks
tend to develop in the original dotted scaly (argillasepic) matrix. (For the birefringence fabric of the fine
mass the terminology of Stoops (1978) is used). These
streaks form a network enclosing dotted scaly areas of
about 50-200 μm diameter. In this way a reticulate

fibrous fabric (more or less equivalent to Brewer's
ortho-bimasepic plasmic fabric) appears. Higher in
the profile the streaks increase in intensity and a slight
decolouration of the fine material is noticeable compared
with the enclosed nascent micropeds which remain red and
become more contrasting. An overall circular fibrous
fabric (no equivalent in Brewer, but comparable to the
ooidsepic of Parfenova and Yarilova, 1972) is now apparent.
Subsequently, microfissures develop in the lighter zones
so that the darker zones become separated from each other
and micropeds are formed. In the ultimate stage the
micropeds tend to develop a more rounded shape, and
the interpedal fissures are transformed into packing
pores. These discrete micropeds retain a dotted scaly
to undifferentiated internal fabric, are surrounded by
a somewhat lighter rim which is birefringent and may
enclose some coarse grains (Fig. 1c and d). Larger cores
of massive, reddish matrix with a dotted scaly fabric are
preserved throughout the profile and remain unaffected.
These may be considered as relicts of the original material
and probably correspond to the pseudonodules identified
by Belgian pedologists in Oxisols from Zaire, and referred
to as variole (Paramananthan and Eswaran, 1980). The
formation of these network-micropeds seems thus to
originate from a change in the fabric of the fine
material; the process is called micro-structuration by
Muller (1977). Activity of the mesofauna may help the
individualisation of these micropeds by pedoturbation.
   According to Chauvel et al. (1978) a micronodular
microstructure can also be formed in this way. In a
clear anisotropic matrix (composed of kaolinite and
iron hydrates) small zones (about 100 μm diameter) with
a darker and more reddish colour are formed as a result
of a different organisation of the iron. These micronodules become discrete when the surrounding matrix tends
to disappear because of its higher mobility as a result
of its partial deferrification and a loose microstructure is formed. These micronodules seem to correspond to the ferritic micropeds of Muller. The relict
micropeds described by the last author must be considered
rather as micronodules inherited from the parent material.
In the upper part of the solum biological activity is
more pronounced and zoogenetic micropeds, packed in

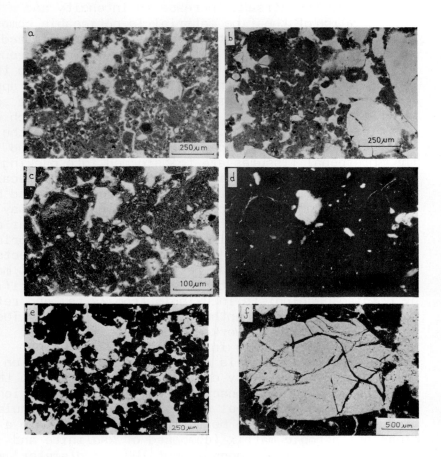

Fig. 1.

aggrotubules or remains of them, become more important.
These micropeds are spheroidal, elipsoidal or irregular,
and have no birefringent rim (Fig. 1e). Complex micropeds
may be of two different origins: (1) small micropeds
(10-15 μm diam.) with a concentric arrangement may form
larger aggregates with a circular fibrous birefringence
fabric. They are assumed to be the result of a microstructuration of a larger microaggregate; (2) biological
microaggregates enclosing one or more other microaggregates.

Summarizing the ideas of ORSTOM pedologists the
conclusion is reached that the process of micro-
aggregation or microstructuration is a general phenomenon
in the B horizons of the red ferrallitic soils. The
process is conditioned by geochemical mechanisms, mainly

Fig. 1. (a) Well developed microped structure
in Oxisol from Rwanda. Different ped sizes
are present, and their rims have mostly a
slightly lighter colour. PPL; (b) Well
developed microped structure in oxic horizon
from Ivory Coast. A better sorting of the peds
is noticed. The composition of the fine mass
is rather homogeneous. PPL; (c, d) Weakly
developed microped structure in Oxisol from
Rwanda. The micropeds are not yet completely
separated. The internal birefringence fabric
is undifferentiated, but a weak circular fibrous
fabric may be observed between the nascent
peds (PPL and XPL); (e) Well sorted zoogenetic
micropeds. Same profile as (c,d) but overlying
horizon. PPL; (f) Runiquartz in Oxisol from
Rwanda. Note the irregular linear alteration
pattern, filled with Fe oxihydrates. PPL.

G. Stoops

by surface interactions between kaolinite and iron compounds. Network-micropeds are characteristic of
homogeneous red materials, whereas ferritic-micropeds
occur in less homogeneously coloured yellowish soils.
Near the surface zoogenetic micropeds may become
important; their genesis is probably more or less
similar to those observed in temperate soils (Bruckert
and Selino,1978).

     The practical significance of this microped
structure has been studied amongst others by Moura
(1970) (mentioned by Buol and Eswaran, 1978) who
showed that the effective volume of the soil that
could be tapped for nutrients is reduced. These soils
will also behave as sandy rather than as clayey materials,
because the micronodules act as pseudosands. The
presence of pseudosands in ferrallitic soils and Oxisols
has long been known by soil scientists working in the
tropics. Micromorphological studies of the sand
fraction, separated after different disaggregation
treatments of the soil, have contributed to a better
evaluation of this phenomenon (Pedro et al., 1976;
Guedez and Langohr, 1978).

The Basic Components of the Oxic Soil Material
     The coarse/fine concept (c/f concept) of Stoops
and Jongerius (1975) is adopted in the following
discussion. The limit between coarse and fine is
situated at approximately 5 µm. This limit has been
chosen in order to include in the fine fraction the
minute opaque, reddish and brownish particles which are
larger than clay size, but generally do not figure as
such in the results of granulometric analyses.

     (a) The coarse material:
     Consists almost solely of stable grains (cp.
definition of the oxic horizon), i.e. mainly quartz,
zircon, tourmaline, rutile and some opaque minerals.
In the least weathered examples small amounts of
muscovite flakes may be present. The silt fraction
is relatively unimportant, and where present in the
mechanical analysis is mainly produced by a fragmentation of sand grains. Verheye and Stoops (1975)

and Chauvel et al. (1978) observed a gradual breakdown of the coarse grains towards the top of the solum. The process seems to be related to a fissuring and corrosion of the quartz grains (Claisse, 1972; Eswaran and Stoops, 1979) and subsequent penetration of iron oxihydrates or gibbsite into the newly formed pores. Such grains with irregular veins of hematite or gibbsite have been called runiquartz by Eswaran et al. (1975). These runiquartz frequently contain forms of iron oxihydrates different from those observed in the surrounding mass (Fig. 1f). On this basis they are features inherited from another environment and can be treated as such in genetic studies.

It was also noticed that the quartz grains of polycyclic Oxisols are more easily fragmented during thin section preparation than quartz grains from less weathered materials.

Sometimes weatherable grains may occur in the solum surrounded and protected by an envelope of iron oxihydrates or gibbsite (Stoops, 1968). In some cases, stable pseudomorphs of more readily weatherable minerals are observed in the solum, e.g. goethite pseudomorphs after garnet (Embrechts and Stoops, in press).

(b) The fine material:
One of the most striking aspects of the fine material of an Oxisol is its homogeneity, which is without doubt the result of a thorough mixing during the pedoplasmation process, and later by bioturbation. The fine material of the oxic horizon is almost exclusively composed of kaolinite and iron oxihydrates in different crystallisation stages : uniform yellowish or reddish staining and/or the presence of goethite or hematite crystallites or droplets (Stoops, 1968). Very finely dispersed gibbsite may also be present. When yellowish, the fine mass is generally limpid, but when reddish a marked cloudiness can be observed in the more evolved types (Bennema et al., 1970). The colour of the fine material is related to the amount and type of iron oxihydrates present. This in turn is a function of the type of parent material, its degree of alteration and the internal drainage

G. Stoops

(Stoops, 1968; Bennema et al., 1970).

Tharmarajan (1979) concluded from a set of experiments that no direct relationship exists between the colour of Oxisols in thin sections and their free iron content, whereas a rather good relationship was found after heating the soil material to 450°C.

One form of the fabric of fine material is its birefringence fabric (Stoops, 1978), i.e. the orientation and distribution pattern of the birefringent domains, and their shape and relative size. The type and development of the birefringence fabric in Oxisols seems to be a function of the composition of the fine material (mainly the type of clay and the amount and type of iron oxihydrates present), the coarse/fine (c/f) ratio, the degree of pedoturbation and the alternation of wetting and drying cycles. Moreover, the observed fabric depends largely on the quality of the thin section, the intensity of the light source and the magnification used. It is therefore normal that quite a range of fabrics have been described in Oxisols by different authors. Some general trends however have been mentioned by several authors. In the most developed Oxisols the fabric seems to be undifferentiated, i.e. no birefringent domains are observed (isotic:sensu Brewer, 1964), (Stoops, 1968; Eswaran, 1972; Bennema et al., 1970), or weakly dotted scaly (asepic to insepic:sensu Brewer). The evolution of the birefringence fabric in relation to the weathering stages of Oxisols in a soil sequence from Malaysia has been described quantitatively by Eswaran and Sys (1976).

In general the birefringent domains in oxic soil materials have only a weak intensity. This is explained both by the low birefringence of kaolinite and the masking effect of the iron oxihydrates absorbed on its surface. Investigations into Oxisols from south-east Asia in our laboratory proved this masking effect : after solution of the iron oxihydrates in the thin section (Bullock et al., 1975) the same fabric exists, but with an increase in intensity. This experiment proves that the absence of birefringent domains is not the result of a masking effect of iron compounds, but that the latter only reduce their

intensity. It was also observed that after heating
the soil samples, whereby all iron oxihydrates are
transformed to hematite, the birefringence fabric may
become more pronounced.

A circular fibrous fabric is commonly observed in
soils with a microped-structure (e.g. Verheye and
Stoops, 1975 : ooidsepic). This fabric is formed by the
juxtaposition of the birefringent rims surrounding the
coalescent microcpeds with an undifferentiated or dotted
scaly internal fabric (cp. discussion on microstructure). More or less well developed fibrous fabrics
may be observed in less developed or less well drained
Oxisols or in their deeper horizons.

Stoops (1968) mentioned the presence of stress-
striotubules in ferralsols from Lower Zaire. These
are thin tubular features with an internal bowlike or
semi-elipsoidal orientation pattern, only visible under
crossed polarisers, in the fine material (Fig. 2a).
Although this is the only mention of this feature in the
literature, investigations at the Ghent State University
show that it is quite common in oxic soil materials
from central Africa and south-east Asia. In some soils
these tubules (c.100 μm diam.) may be so numerous
that they determine its birefringence fabric (Fig. 2b).
The name crescent-fibrous is proposed for this fabric.
It is clearly of biological origin.

Cagauan and Uehara (1965) found a relationship
between aggregate stability in some latosols from
Hawaii and the intensity of the birefringence fabric.
Many more investigations on this topic are necessary.

(c) c/f related distributions (c/f R.D.);   Stoops
and Jongerius, (1975):
Since, by definition, a rather large amount of clay
is present in the oxic horizons, practically only a
porphyric c/f R.D. is observed. Depending on the c/f ratio
this may be close to open spaced. In oxic sandy materials
an enaulic c/f R.D. may be observed. Eswaran (1978)
mentions an agglutinic SRDP as typical of most oxic
horizons. This pattern is considered here as a microped
structure, and intrapedally a porphyric c/f R.D. is
observed.

Pedofeatures
    In situ formed features are relatively rare in oxic
horizons for two reasons : the physicochemical immobility
and stability of the constituent material, and the fact
that homogenising processes are very active in these
soils. Nevertheless, a few features are commonly
observed in oxic horizons, although they are not
diagnostic.

    (a) Pedotubules:
    Pedotubules are frequently observed in most oxic
horizons. Aggrotubules may contribute to the formation
of the characteristic microped structure. Striotubules
have been observed as a main fabric characteristic of the
casing of termite mounds on Oxisols (Stoops, 1965). It
is therefore reasonable to expect that some of the
striotubules observed in oxic horizons may be related to
termite activity. Granotubules seem to be very rare.

Fig. 2. (a)Crescent-fibrous birefringence fabric :
         detail of the wormlike birefringence pattern.
         Haplorthox, Malaysia (circular polarised light;
         (b) Crescent-fibrous birefringence pattern :
         general aspect. Ferralsol, Zaire. XPL; (c)
         Fe oxihydrate deposits (plinthite) in Oxisol
         from Zaire. Reddish flecks occur in a greyish
         mass. PPL; (d) detail of (c). Note the formation
         of irregular and diffuse nodules and subcutans;
         (e) Goethite nodule in Oxisol from The Cameroons.
         The well preserved boxwork fabric indicates a
         pseudomorph after garnet. PPL; (f) Gibbsite
         cutan around quartz grains, which is typical
         for strongly weathered material. Oxisol from
         Zaire. XPL.

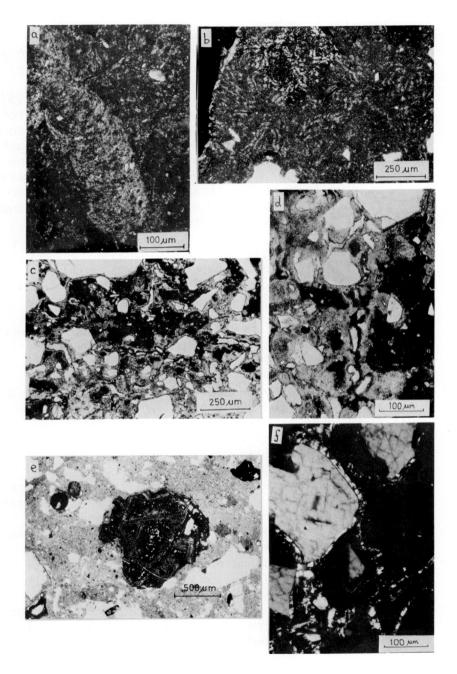

Fig. 2.

(b) Clay cutans:

According to Soil Taxonomy (Soil Survey Staff, 1975), no more than 1% clay cutans may be present in an oxic horizon. From a statistical viewpoint, however, this limit seems questionable (Hill, 1970). Larger amounts have been described by Benayas and Refega (1974) and Chauvel (1977), but it is questionable whether these soils can still be classified as Oxisols. Illuviation clay cutans are frequently observed in the deeper part of Oxisols (e.g. also in the saprolite). Thin cutans near the surface may be a result of local (e.g. lateral) clay translocation, rather than a vertical eluviation and illuviation process, and are related to drying and wetting cycles of the soil (Stoops, 1968).

(c) Iron oxihydrate segregations and nodules:

Iron oxihydrate segregations frequently occur in the less well drained Oxisols. These segregations appear in general as irregular and rather diffuse brownish or reddish flecks and/or as neo-cutans around pores and grains. A more extensive development of these features gives rise to the formation of a plinthite layer. Micromorphologically this is characterised by a juxtaposition of greyish and reddish zones (Fig. 2c and d). These reddish mottles or streaks have in general an undifferentiated birefringence fabric, whereas the greyish zones may display a scaly or even a weakly fibrous fabric. A more detailed discussion of the plinthite fabric is beyond the scope of this paper.

A common feature in many Oxisols is the presence of hard laterite nodules, which are frequently concentrated in a stone-line. Part are allochthonous and mainly erosion products of older laterite levels. Others may be considered as disorthic, i.e. they are formed within the profile but have undergone some relative displacement or translocation (Stoops, 1967). Both types have sharp boundaries and are at least partly rounded. Frequently they are also fragmented, as deduced from the abrupt interruption of some continuous internal features (e.g. enclosed cutans).

    Both allochthonous and disorthic nodules may display
different types of internal fabric depending upon the
degree of evolution of the material that has been iron
impregnated. In fact, the degree of iron impregnation
can indicate the stage reached in the general evolution
of the material. The internal fabric of the nodules may
reflect the following main stages : (1) incomplete
saprolite (enclosing some weatherable minerals such as
biotite or muscovite); (2) complete saprolite (only
pseudomorphs after primary minerals are present, but the
original rock fabric is preserved); (3) altered saprolite
(the rock fabric is partly altered); (4) soil (soil
fabric). Several subtypes can be distinguished in the
latter according to the type of soil fabric present (e.g.
an argillic fabric with ferruginised clay cutans). The
saprolitic types may be subdivided according to the
original rock fabric (e.g. granitic, schistose,
doleritic). In addition, pseudomorphs after different
mineral species (viz. garnet, olivine) may occur as
nodules (Fig. 2e).
    The nodules may be subdivided also on the basis of
mineralogical composition into goethitic and hematitic;
several transitions and combinations are also possible.

    (d) Gibbsitic features:
    Although gibbsite is a common component of the clay
fraction of many Oxisols, gibbsitic features are not
abundant in soil thin sections. Two main types may be
distinguished : inherited features and new formations.
The inherited features are mainly gibbsite pseudomorphs
after feldspars, or even completely transformed rock
fragments which have escaped homogenisation. The new
formation of gibbsite in Oxisols has been discussed by
Eswaran et al. (1977). The mineral can occur as
nodules but frequently forms cutans on fissures or
channels, or around coarse grains (Fig. 2f). Well
developed gibbsitic features are observed in the
Gibbsiorthox (Eswaran, 1978). These have formed by
direct crystallisation in situ from the soil solution.
The mode of formation discussed by Boulange et al.
(1975) does not seem valid for these soils, as no
illuviation of kaolinitic clay takes place. Moreover,

G. Stoops

a different internal fabric of the features would
result.

    The results of a recent investigation (Embrechts and
Stoops, in press) point to an unexpected phenomenon :
gibbsite present in features inherited from the saprolite
tends to dissolve in some oxic horizons. A resilification
of gibbsite (mostly to kaolinite) has been described by
several authors. In these cases, however, morphologically
only a congruent dissolution can be observed.

    (e) Siliceous features:
    Although the new formation of siliceous features
seems rather contradictory in the context of the formation
of Oxisols, it has been mentioned in the literature
repeatedly.
    Laruelle (1956) reports the presence of an authi-
genic fine quartz powder as "an alteration product from
once mobile silica" in the most weathered soils of the
Garamba Park (Zaire). Eswaran (1972) mentions the
presence of newly formed quartz nodules and even
quartz cutans in Oxisols on basalt in Nicaragua.
New formation of bipyramidal quartz-microcrystals in the
stone-line of some ferralsols in Zaire has been
observed by Eggoroff (pers. comm.).

DISCUSSION
    According to Buol and Eswaran (1978) and Eswaran
(1978) it is not possible to identify an oxic horizon
on the basis of one micromorphological characteristic.
Indeed, all the features described in this paper are
typical for oxic soil materials, but none is unique :
each can occur also in other soil materials, but only
in Oxisols do they all occur together. So only a
combination of several characteristics will be
diagnostic and in general the more they are developed,
the more typical micromorphological features will be
present and the more they will be pronounced.
    Buol and Eswaran (1978) introduced the concept of
the oxic syndrome, i.e. a set of features characterizing
the oxic horizon by their diagnostic presence or absence.
According to these authors the oxic horizon (1) may
have any NRDP except granic or phyric (note : granic and
phyric NRDP correspond to monic c/f R.D. respectively

of sand and silt grains); (2) has a poorly expressed
plasmic fabric-insepic, argillasepic or isotic;
(3) has only traces of weatherable minerals comprising
the grains; (4) has only traces of translocated clay
unless present as papules; (5) may have an agglutinic
SRDP.

   According to our experience in Ghent and data from
literature, the following features can be considered as
micromorphological characteristics for an oxic soil
material :

1. Absence or very low content (traces) of weatherable
   minerals in the sand fraction as well as in the silt
   fraction, unless protected by an envelope of sesquioxides;
2. Very low silt content (except for fine iron oxihydrate
   particles), unless present in inherited sesquioxidic
   nodules;
3. Absence or very low amount (< 1 %) of illuviated
   clay, recognisable as clay cutans or clay papules.
   These three characteristics are necessary for any
   soil material to be defined as oxic, as they are part
   of its definition. Other possible characteristics are :
4. Presence of a microped structure : network and
   ferritic micropeds are, in particular, good indicators;
   zoogenetic micropeds are less characteristic.
5. An undifferentiated or weakly developed scaly birefringence fabric; in the case of a network-microped
   structure a weak circular fibrous fabric may be
   present; a weak crescent-fibrous fabric may occur in
   more compact materials.
6. Lateritic gravel may be present, usually concentrated
   in a stone-line.
   From the above it may be concluded that it is possible
to identify an oxic soil material by micromorphological
techniques, but it does not follow that an oxic horizon can
be characterised on the basis of its micromorphological
aspects only, as other factors (field characteristics,
physico-chemical data) should be taken into consideration.
   On the basis of this review it is evident that
micromorphology has played, and still has to play, an
important role in the study of Oxisols and oxic horizons.

Micromorphology has already contributed a great deal to a
better understanding of the genesis of these materials and
their characteristics (e.g. pedoplasmation, structuration)
and to the relation that exists between the different soil
types and their interdependency (e.g. Chauvel, 1977). As
mentioned by Eswaran et al. (1979), micromorphology also
has a role to play in defining lower taxa of the soil
classification.

     Compared with other soil orders, relatively little
has been published on the micromorphology of Oxisols.
Moreover, interpretation of the existing literature is
difficult for several reasons : (1) only a few good and
complete objective descriptions have been published; in
many cases only a discussion of a given characteristic
is given; (2) the micromorphological terminology used
is not always sufficiently well defined; (3) for many
soils the exact classification according to Soil
Taxonomy could not be made as insufficient analytical
or field data are presented; frequently no detailed
classification is given by the author(s), not even
within their own national classification system.

     Significantly, too, no work seems to have been
published on the effects of cultural practices on the
micromorphological properties of oxic horizons. It is the
opinion of the author that a large field of practical
investigation remains open.

REFERENCES

Beaudou, A.G. 1972. Expression micromorphologique de la
     microagrégation et de l'illuviation dans certains
     horizons de sols ferrallitiques centrafricains et
     dans les sols hydromorphes associés. Cahiers ORSTOM,
     série Pédologie, 10, 357-372.

Beaudou, A.G. and Chatelin, Y. 1979. La pédoplasmation
     dans certains sols ferrallitiques rouges de savane
     en Afrique Centrale. Cahiers ORSTOM, série Pédologie,
     17, 3-8.

Benayas, J. and Guerra Refega, A. 1974. Aplicacion de
     la micromorfologia a la clasificacion de algunos
     suelos ferraliticos de Angola. Ans. Edaf. y Agrobiol.,
     33, 283-294.

Bennema, J., Jongerius, A. and Lemos, R. 1970. Micro-
    morphology of some oxic and argillic horizons in
    south Brazil in relation to weathering sequences.
    Geoderma, 4, 333-355.

Boulange, B., Paquet, H. and Bocquier, G. 1975. La rôle
    de l'argile dans la migration et l'accumulation de
    l'alumine de certaines bauxites tropicales. Comptes
    Rendus Hebdomadaires des Séances de l'Académie des
    Sciences. Paris, t. 280, sér. D, 2183-2186.

Brewer, R. 1964. Fabric and Mineral Analysis of Soils.
    John Wiley & Sons, New York, 470 pp.

Bruckert, S. and Selino, D. 1978. Mise en évidence de
    l'origine biologique ou chimique des structures
    microagrégées foisonnantes des sols bruns ocreux.
    Pédologie, 28, 46-59.

Bullock, P., Loveland, P.J. and Murphy, C.P. 1975. A
    technique for selective solution of iron oxides in
    thin sections of soil. J. Soil Sci., 26, 247-249.

Buol, S.W. and Eswaran, H. 1978. The micromorphology of
    Oxisols. In : M. Delgado (Ed), Micromorfologia de
    Suelos. Universidad de Granada, 325-347.

Cagauan, B. and Uehara, G. 1965. Soil anisotropy and
    its relation to aggregate stability. Soil Sci.
    Soc. Amer. Proc., 29, 198-200.

Chauvel, A. 1977. Recherches sur la transformation des
    sols ferrallitiques dans la zone tropicale à saisons
    contrastées. Travaux et documents de l'ORSTOM, no.
    62, 532 pp.

Chauvel, A., Bocquier, G. and Pedro, G. 1978. La
    stabilité et la transformation de la microstructure
    des sols rouges ferrallitiques de Casamance
    (Sénégal). Analyse microscopique et données
    experimentales. In : M. Delgado (Ed), Micro-
    morfologia de Suelos. Universidad de Granada,
    779-813.

Claisse, G. 1972. Etude sur la solubilisation du quartz
    en voie d'altération. Cahiers ORSTOM, série
    Pédologie, 10, 97-122.

Condado, J.L.A. 1969. Micropedologia de alguns dos
    mais representativos solos de Angola. Memorias
    da Junta de Investigaçoes do Ultramer, 59, 145 pp.

Embrechts, J. and Stoops, G. Micromorphological aspects of garnet weathering in a humid tropical environment. (in press).

Eswaran, H. 1972. Micromorphological indicators of pedogenesis in some tropical soils derived from basalts from Nicaragua. Geoderma, 7, 15-31.

Eswaran, H. 1978. Micromorphology of Oxisols. International Soil Classification Workshop. Malaysia-Thailand. August 28th to September 9th 1978, 9 pp.

Eswaran, H., Sys, C. and Sousa, E.C. 1975. Plasma infusion - A pedological process of significance in the humid tropics. Ans. Edaf. y Agrobiol., 34, 665-674.

Eswaran, H. and Baños, C. 1976. Related distribution patterns in soils and their significance. Ans. Edaf. y Agrobiol., 35, 33-45.

Eswaran, H. and Sys, C. 1976. Micromorphological and mineralogical properties of the Quoin Hill toposequence (Malaysia). Pédologie, 26, 152-167.

Eswaran, H., Stoops, G and Sys, C. 1977. The micromorphology of gibbsite forms in soils. J. Soil Sci., 28, 136-143.

Eswaran, H. and Stoops, G. 1979. Surface textures of quartz in tropical soils. Soil Sci. Soc. Amer. J., 43, 420-424.

Eswaran, H., Van Wambeke, A. and Beinroth, F.H. 1979. A study of some highly weathered soils of Puerto Rico. Micromorphological properties. Pédologie, 29, 139-162.

Flach, K.W., Cady, J.G. and Nettleton, W.D. 1968. Pedogenic alteration of highly weathered parent materials. Transactions of the 9th Int. Congr. Soil Sci., IV, 343-351.

Folster, H. 1964. Die Pedi-sedimente der Südsudanischen Pediplane. Herkunft und Bodenbildung. Pédologie, 14, 64-84.

Folster, H. 1971. Ferrallitische Böden aus sauren metamorphen Gesteinen in den feuchten und wechselfeuchten Tropen Afrikas. Göttinger bodenkundliche Berichte 20, 231 pp.

Guedez, J.E. and Langohr, R. 1978. Some characteristics of
    pseudosilts in a soil toposequence of the Llanos
    orientales (Venezuela). Pédologie, 28, 118-131.
Harrassowitz, H. 1930. Böden der tropischen Regionen.
    In : E. Blanck (Ed), Handbuch der Bodenlehre Bd.
    3. Springer, Berlin.
Harrison, J.B. 1933. The katamorphism of igneous rocks
    under humid tropical conditions. Imp. Bur. Soil
    Science, Harpenden, England. 79 pp.
Hill, I.D. 1970. Quantitative micromorphological evidence
    of clay movement. In : D.A. Osmond and P. Bullock
    (Ed), Micromorphological Techniques and Applications.
    Soil Survey Tech. Monogr. 2. Harpenden, England,
    33-42.
Kubiena, W.L. 1948. Entwicklungslehre des Bodens.
    Springer-Verlag, Wien. 215 pp.
Laruelle, J. 1956. Quelques aspects de la microstructure
    des sols du nord-est du Congo Belge. Pédologie,
    6, 38-58.
Muller, J.P. 1977. Microstructuration des structichrons
    rouges ferrallitiques, à l'amont des modelés
    convexes (Centre Cameroun). Aspects morphologiques.
    Cahiers ORSTOM, série Pédologie, 15, 239-258.
Paramananthan, S. and Eswaran, H. 1980. Morphological
    Properties of Oxisols. In : B.K.G. Theng (Ed),
    Soils with Variable Charge. New Zealand Soc.
    Soil Sci., 35-43.
Parfenova, E.I. and Yarilova, E.A. 1972. Schemes of soil
    fabric components. In : St. Kowalinski (Ed),
    Soil Micromorphology. Warsaw, Poland, 39-55.
Pedro, G., Chauvel, A. and Melfi, A.J. 1976. Recherches
    sur la constitution et la genèse des Terra Roxa
    Estructurada du Brésil. Ann. Agron., 27, 265-294.
Romashkevich, A.E. 1964. Micromorphological indications
    of the processes associated with the formation of
    the krasnozems (red earths) and the red-coloured
    crust of weathering in the Transcaucasus. In : A.
    Jongerius(Ed), Soil Micromorphology. Elsevier,
    Amsterdam, 261-268.
Romashkevich, A. 1974. Soils and weathering crusts of
    the humid subtropical West Georgia (USSR). Akademia
    Nauk USSR, Int. Congr. Soil Scientists,Moscow 1974,
    217 pp.

Soil Survey Staff, 1975. Soil Taxonomy. A basic system of
    soil classiciation for making and interpreting soil
    surveys. Agriculture Handbook. No.436. Soil Con-
    servation Service, U.S.D.A., Washington, D.C.
Stoops, G. 1964. Application of some pedological methods
    to the analysis of termite mounds. In : Etudes
    sur les Termites Africains. Edition Université
    Lovanium, 379-398.
Stoops, G. 1967. La profil d'altération au Bas-Congo
    (Kinshasa). Sa description et sa genèse. Pédologie,
    17, 60-105.
Stoops, G. 1968. Micromorphology of some characteristic
    soils of the Lower Congo (Kinshasa). Pédologie, 18,
    110-149.
Stoops, G. 1978. Provisional Notes on Micropedology.
    State University Ghent, 120 pp. (stencil).
Stoops, G. and Jongerius, A. 1975. Proposal for a
    micromorphological classification of soil materials.
    I. A classification of the related distributions
    of fine and coarse particles. Geoderma, 13,
    189-199.
Tharmarajan, M. 1979. Mineralogical characterisation
    of two toposequences from Sarawak, Malaysia. MSc
    Thesis, State University Ghent, 95 pp.
Vageler, P. 1930. Grundriss der tropischen und
    subtropischen Bodenkunde, Berlin : Verlaggesellsch.
    f. Ackerbau.
Verheye, W. and Stoops, G. 1975. Nature and evolution of
    soils developed on the granite complex in the
    subhumid tropics (Ivory Coast). II. Micromorphology
    and mineralogy. Pédologie, 25, 40-55.

MICROMORPHOLOGICAL EVIDENCE OF TURBATION IN VERTISOLS AND SOILS IN VERTIC SUBGROUPS

W.D. Nettleton[1], F.F. Peterson[2] and G. Borst[3]

[1] USDA, Soil Conservation Service, National Soil Survey Laboratory, Lincoln, Nebraska, U.S.A.
[2] University of Nevada, Reno, Nevada, U.S.A.
[3] USDA, Soil Conservation Service, Fallbrook, California, U.S.A. (Retired).

ABSTRACT
The eight Vertisols and six soils in vertic subgroups of this study were sampled in 13 counties across the United States. The soils are clayey, except for the surface horizons of some of the vertic subgroups. They have deep, wide cracks that open to the soil surface or to thin, lighter textured surface horizons at some time during the year, and they have high bulk densities between the cracks. The soils in vertic subgroups do not have intersecting slickensides or other direct evidence of turbation. Except for the surface horizons of those soils in vertic subgroups which have argillic horizons, the chemical and physical properties of the Vertisols and soils in vertic subgroups cannot be differentiated. There are, however, differences in their morphology and micromorphology. The properties of the soils in the vertic subgroups are the most variable. The Vertic Torrifluvents have silasepic plasmic fabric, the Vertic Hapludalfs, omnisepic. One of the Vertic Hapludalfs has a few slickensides, and in this horizon the plasmic fabric is ma-omnisepic. Most of the Vertisols have skelsepic plasmic fabric in upper parts and masepic, or an intergrade toward that kind of fabric, in lower parts. The Vertisols also differ from those soils in vertic subgroups in having silt and sand size aggregates of organic matter mixed throughout the plasma in upper horizons. The turbation in Vertisols probably results from differential wetting of dry soil. The Vertisols wet deeply and are churned or mixed, and surface horizons erode, filling the cracks. Water passes into the cracks causing the montmorillonitic clay soil at the bottom to wet first. In many instances, the expansion of this wet

soil lifts the column of dense drier soil between the
cracks.

INTRODUCTION
     The clayey soils examined in this paper have deep
wide cracks, $\geqslant$ 1 cm at 50 cm depth at times in most years,
and have high bulk densities between the cracks (Soil
Survey Staff, 1975). These soils were rendzinas, gray
brown podzolics or planosols in the 1938 classification
(Baldwin et al., 1938), and regosols, grumusols,
planosols, gray brown podzolics, or alluvial soils after
1951 (Thorp and Smith, 1949). Some of them are now
classified as Vertisols. Vertisols are mineral soils
"that have 30% or more clay in all horizons down to a
depth of 50 cm or more after the surface soil to a depth
of 18 cm has been mixed, that at some period in most years
have cracks that are open to the surface or to the base
of a plough layer or surface crust, and that are at
least 1 cm wide at a depth of 50 cm unless the soil is
irrigated...." (Soil Survey Staff, 1975). Among the
soils studied, the Vertisols have one or more of the
following: gilgai, intersecting slickensides, wedge-
shaped aggregates. The other soils included in this
paper have the cracks required of Vertisols but lack the
other features. These soils are referred to herein as
vertic subgroups.
     Vertisols and vertic subgroups (Soil Survey Staff,
1975) are believed to have had their distinctive properties formed by mixing processes resulting from changes
in moisture content. Hole (1961) uses the term
pedoturbation' as a synonym for soil mixing. He has
termed the process of mixing of montmorillonitic clay-
rich soils by expansion following contraction and the
filling of cracks as 'argillipedoturbation'.
     Knight (1980) has reviewed the literature on kinds of
gilgai and agents and mechanisms proposed for formation of
gilgai soils. In addition, he has made detailed
structural analysis of gilgai in Victoria, Australia.
Kormonik and Zeitlin (1970), and Katti et al. (1969) found
that the horizontal stress component in swelling soils
exceeds the vertical stress by a factor of 4 or more.
They reported a net vertical stress due to swelling of
2.1 kg/cm^2 in the 75 to 100 cm depth range.

In this paper soil micromorphology is used as a
tool to examine the strain features produced by changes in
soil moisture content.   Such strain in fine textured
soils is believed to result from vertical cracking in
the uppermost clay horizons and from horizontal cracking
with increasing depth in fine textured soils (Sleeman,
1963).   In thin section, the analysis of planar void
patterns can be expected to be an expression of deposi-
tional and/or stress-strain history of the particular
soils (Lafeber, 1965).   In the light of this, stress
should be related to birefringence of clay, porosity
in the form of planar voids and frequency of slickensides.   The plotting methods of Lafeber (1965) were not
used by us since the orientation of most of the sections
is not known.
     Gallavan and Greene-Kelly (1974) made some model
experiments that suggest that simple wetting and drying
of unconfined soils will not produce re-organisation of
clay domains.   Earlier, Towner (1961) had shown that
shrinkage of a saturated clay soil on drying can be
viewed as the compression of the soil fabric by an all
round pressure that is quantitatively equivalent to the
internal soil water suction. Greene-Kelly and Mackney
(1970) were unable to detect any change either in size
or number of domains or in preferred orientation when
remoulded clay soils were dried, even if drying was
deliberately uneven.   In applying these results to soils,
it should be kept in mind that the results were obtained
on pure clay systems.   Introducing silt, sand, gravel,
or pedological features into these systems may change
the results.
     Although, for describing soils, micromorphology had
gained wide acceptance by 1969, De Vos and Virgo (1969)
were able to cite only the work of Jongerius and Bonfils
(1964) on the micromorphology of Grumusols and that of
Rode et al. (1960) on the micromorphology of dark soils
of south Russia.   They reported that clay domains within
the matrix show weak bimasepic orientation while lower
in the profiles, there is skelsepic orientation around
the larger glaebules.   A vosepic pattern was observed
along the surfaces of the majority of the straight-sided
voids indicating that these represent slickensided faces.

Bellinfante et al. (1974) found an agglomeroplasmic
related distribution and asepic plasmic fabric in the Ap
horizons of Vertisols in southern Spain. The horizons
below were porphyroskelic with insepic, masepic and
lattisepic plasmic fabric. They found that chambers
exceeded the number of shrinkage cracks in the soils.
Nemeçek (1975), in a summary of micromorphological work
on Czechoslovakian soils, found Vertisols to have masepic
or omnisepic plasmic fabric.
      In this paper, the micromorphology of Vertisols and
vertic subgroups studied in the United States together
with the processes responsible for their formation
is examined.

METHODS AND MATERIALS
      The methods are described in U.S. Soil Conservation
Service (1972). The codes listed with the data (Table 1)
are the ones used to list the methods in that publication.
For the most part, thin sections were prepared by the
methods of Innes and Pluth (1970). The micromorphology
terms are those of Brewer (1964).
      The soils are listed in Table 2; they come from the
states of California, Nevada, New Mexico, Virginia,
North Carolina, and South Carolina. Xeric, torric,
ustic, and udic moisture regimes are included. The
torric environment studied approaches ustic.

RESULTS
      The physical and chemical properties of the horizons
of the vertic subgroups (without the A horizons of the
Vertic Hapludalfs) and of the horizons of the Vertisols
appear to have come from the same population (Table 1).
The sand is mostly one-third or less of the < 2 mm soil
by weight. Clay amounts to one-third to two-thirds. Clay
content of the surface horizon of one of the Vertisols is
less than 30%, but all the other horizons of the Vertisols
are over 30. Atterberg limits, linear extensibility,
and cation exchange capacity values of both groups of
soils are high as would be expected of soils with
montmorillonitic clay mineralogy. All the samples
disperse well.

Table 1. Summary of physical and chemical properties of the soils.

Statistics	Sand IB1B 3A1 (%)	Clay IB1B 3A1 (%)	Liquid limit 4F1 (%)	Plastic limit 4F2 (%)	15-bar water 4B2 (%)	D_b O.D. 4A1H (g/cc)	LE 41D (%)	CEC 5A6A (me/100g)
Vertic Torrifluvents[1] and Vertic Hapludalfs[2] (9 horizons)			*					
Mean	12.6	56.0	67.3	40.3	23.9	1.8	11.1	38.7
Maximum	23.7	69.2	76.0	47.0	27.7	2.00	15.0	49.0
Minimum	4.9	31.5	40.0	29.0	16.9	1.53	4.0	19.8
Std. Dev.	7.3	12.7	11.1	6.7	4.1	0.14	4.1	8.8
Chromoxererts, Pelloxererts, and Chromusterts (26 horizons)								
Mean	15.7	46.4	n.a.	n.a.	20.8	1.75	8.7	40.2
Maximum	35.1	68.5	n.a.	n.a.	39.7	1.95	15.9	58.1
Minimum	4.8	27.1	n.a.	n.a.	12.2	1.31	4.1	24.6
Std. Dev.	7.3	8.2	n.a.	n.a.	5.9	0.16	2.1	9.5

1 Both A and C horizons are included.

2 Includes only B2t horizons.

* Atterberg limits were not determined for these samples.

W.D. Nettleton et al.

Table 2.   Series name, location, and classification
            of the pedons.

Series	Location	Classification*
Imperial	Imperial County, CA	Vertic Torrifluvent, fine, montmorillonitic (calcareous), hyperthermic
Gadsen	Imperial County, CA	Vertic Torrifluvent, fine, montmorillonitic (calcareous), hyperthermic
Iredell (taxadjunct)	Louisa County, VA	Vertic Hapludalf, fine, montmorillonitic, thermic
Iredell (taxadjunct)	Chester County, SC	Vertic Hapludalf, fine, montmorillonitic, thermic
White Store (taxadjunct)	Durham County, NC	Vertic Hapludalf, fine, montmorillonitic, thermic
White Store (taxadjunct)	Goochland County, VA	Vertic Hapludalf, fine, mixed, thermic
Verhalen	Grant County, NM	Mollic Torert, fine, montmorillonitic, thermic
Day	Ada County, ID	Typic Chromoxerert, very fine, montmorillonitic, mesic
Alo	Orange County, CA	Typic Chromoxerert, fine, montmorillonitic, thermic
Capay	Riverside County, CA	Typic Chromoxerert, fine, montmorillonitic, thermic
Karlo	Washoe County, NV	Typic Chromoxerert, very fine, montmorillonitic, frigic
Diablo	Monterey County, CA	Chromic Pelloxerert, fine, montmorillonitic, thermic
Bosanko	San Diego County, CA	Chromic Pelloxerert, fine, montmorillonitic, thermic
Bickerdyke	Carter County, MT	Udorthentic Chromustert, very fine, montmorillonitic, frigid

* Soil Survey Staff, 1975.

All but one of the horizons of the soils in vertic
subgroups have omnisepic plasmic fabrics (Fig. 1).
Most of the horizons have linear extensibilities greater
than 9%. One horizon has masepic plasmic fabric.

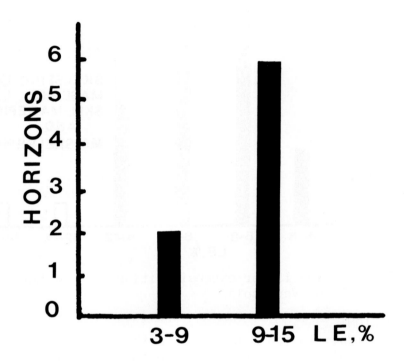

Fig. 1. Linear extensibility (LE) of horizons of
        the soils in vertic subgroups with omnisepic
        fabric. One horizon in the 3 to 9 percent LE
        class had mosepic plasmic fabric. Surface
        horizons of the soils having argillic horizons
        were not included.

Most of the Vertisols have skelsepic plasmic fabric
(Fig. 2), with also small areas of masepic plasmic fabric
in their lower horizons. Three horizons of the soils
studied were masepic at depth. All of the Vertisols have
silt and clay-size pieces of organic matter or organic
and mineral matter mixed throughout their upper horizons.

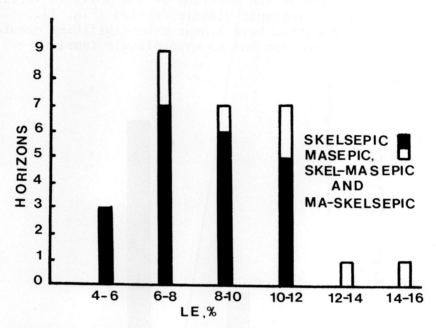

Fig. 2.   The linear extensibilities (LE) of horizons
         of the Vertisols.

DISCUSSION
     The data in Table 1 show that the vertic subgroups and
the Vertisols cannot be distinguished by physical and
chemical properties alone.   There are, however, striking
differences in their morphology and micromorphology.

The Vertic Subgroups
     Vertic Torrifluvents represent the minimum strain
one would expect to find in vertic subgroups.   The two
series of Vertic Torrifluvents studied, the Imperial and
Gadsen soils (Table 2), have large potential to shrink
and swell because of their high content of montmorillonitic clay.   They have a dense porphyroskelic related
distribution (Fig. 3a) but show no plasmic fabric evidence
of stress (Fig. 3b).   The fine stratification (0.2 to 0.5
mm) (Fig.4a, b) of the parent material of the Imperial
soils has even survived in the Ap and Cl horizons.   This
is true even though the soils have cracks a cm or more

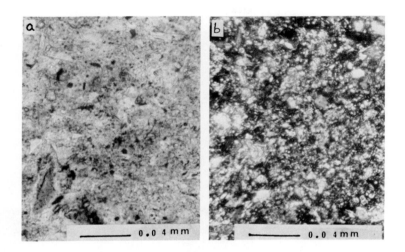

Fig. 3. The relatively dense related distribution
         (porphyroskelic) of the Cl horizon (33-71 cm)
         of the Imperial pedon (Vertic Torrifluvent) from
         Imperial County, California. (a) PPL, (b) XPL –
         note lack of evidence of strain.

wide that extend from the surface to 50 cm or more depth.
These soils are usually dry. Apparently, they have not
been through enough wetting-drying cycles to reach their
potential for turbation.
     Two pedons of each of two series of Vertic Hapludalfs
were studied. These soils have clay-rich B horizons and
densely packed porphyroskelic related distributions
(Fig. 4c). The plasmic fabric (omnisepic)(Fig. 4d,e) of
these soils shows some evidence of strain. Orientation
of the fabric forms short areas of various dimensions.
These soils are usually moist but dry enough to crack to
the base of the surface horizon for short periods in most
years. They lack the silt and clay-size pieces of organic
matter or organic and mineral matter that are found in
Vertisols. Apparently, mixing in these soils is not
intense enough to move parts of the A horizon into the
underlying horizons.

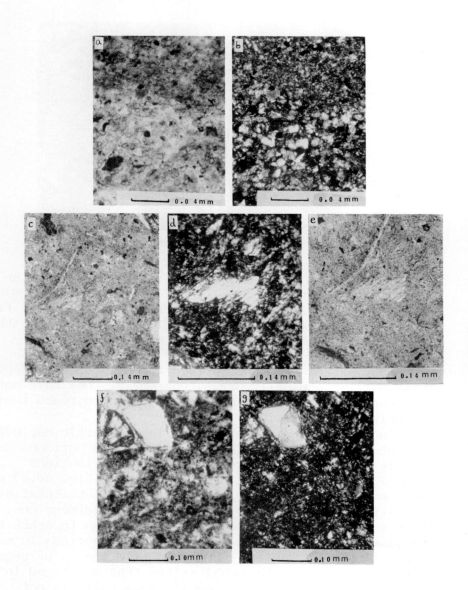

Fig. 4.

The Torrerts
      The Mollic Torrerts (Verhalen series) have a very
dense porphyroskelic related distribution also (Fig. 4
f,g). Clays are stress oriented around sand-size
particles (skelsepic plasmic fabric) (Fig. 5a,b). The
surface horizons of these soils have granular structure
in which the granules are 0.1 to 1 mm in cross section
and very dense. Most are 0.2 to about 0.4 mm in cross
section and have a porphyroskelic related distribution
(Fig. 5c). The soils receive several light rain showers
in most summers.
      One of the authors (F.F.P.), in unpublished work on
the Diablo soils, was able to change their surface soil
structure from massive to granular and then back to
massive by artificially wetting surface horizons of
this Pelloxerert. High tension wetting-drying cycles
resulted in a granular or crumbly structure. Wetting
samples to near saturation or puddling followed by drying,
produced a massive or crusty structure. Hence, one would
expect the granular structure of the Verhalen, and
other soils in a low annual precipitation ustic or near
ustic moisture regime to result from frequent wetting
at relatively high tension.

      Fig. 4. (a,b) Stratification in the Cl horizon (33-71
            cm) of the Imperial pedon (Vertic Torrifluvent)
            from Imperial County, California. The strata are
            0.2 to 0.5 mm thick.(a) PPL and (b) XPL; (c)
            The densely packed basic related distribution
            (porphyroskelic) of the B21t horizon (28-43 cm)
            of the Iredell pedon from Virginia. PPL; (d,e)
            Part of field of view in (c) showing detail
            of omnisepic plasmic fabric and weathered
            plagioclase (centre). Dark grains in plain light
            (e) are mostly weathered biotite; (f,g) The very
            dense related distribution (porphyroskelic) of
            the Al2 horizon (3-18 cm) of Verhalen pedon
            17-23. Note the stress oriented clay around the
            feldspar grain in the upper left-hand corner of
            (g), between crossed polarisers.

452   W.D. Nettleton et al.

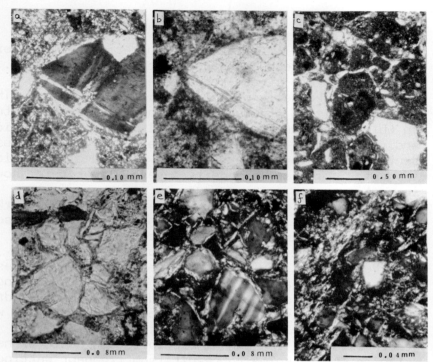

Fig. 5. (a,b) Skelsepic plasmic fabric of the A12 horizon (3-18 cm) of Verhalen pedon 17-23. Note the
stress oriented clay around the plagioclase sand
grain between crossed polarisers (a) and incomplete mixing of organic matter and the plasma in
plain light (b); (c) Granular structure of the
A12 horizon (3-8 cm) of Verhalen pedon 17-21. PPL;
(d) Porphyroskelic related distribution of the A11
horizon (0-10 cm) of Capay pedon 33-2. This horizon has only 27% clay and borders on agglomeroplasmic related distribution. Most of the sand-
size grains are plagioclase; two at the top of
figure are hornblende. PPL; (e) Same field as
(d) except it has been rotated counter-clockwise
about 45º. Note the stress oriented clays
(skelsepic plasmic fabric) around sand grains.
XPL; (f) Masepic plasmic fabric in the A12 horizon (10-25 cm) of Capay pedon 33-2. The fabric
is weakly developed because of the mixture of
silt-size aggregates of organic matter and clay.
XPL.

The Chromoxerts

Chromoxerts of five soil series of California, Idaho and Nevada were studied. All contain silt and clay-size pieces of organic matter or organic and mineral matter in upper horizons. The basic structure of each of these Chromoxerts is relatively dense in upper parts, and becomes denser with depth. All have a porphyroskelic related distribution (Fig. 5d). Clays are stress oriented around sand-size particles (skelsepic plasmic fabric) (Fig. 5e) in the upper parts. There are few slickensides in these parts and sand-size particles are an important component. In the middle and lower parts, there are some microslickensides (masepic plasmic fabric) (Fig. 5f). In most of the soils formed in granites, or in sediments derived from them, skelsepic is the dominant plasmic fabric, but in the lower parts there are a few masepic areas associated with the slickensides.

The Pelloxererts

Pelloxererts from two series, Bosanko and Diablo, were studied. The Bosanko soils are formed in granitic rocks; Diablo, in marine shales. They have the same patterns of structure and fabric noted for the Chromoxerts; i.e., porphyroskelic related distributions (Fig. 6a) in upper horizons, becoming denser with depth and with skelsepic plasmic fabric grading to masepic (Fig. 6b) with increasing depth. Again, the pieces of silt and clay-size organic matter or organic and mineral matter somewhat mask the orientation of the plasma.

Chromusterts

One Chromustert, Bickerdyke, formed in marine shales was included. Formerly, this soil was classified as a Borollic Vertic Camborthid because it was in a frigid family. This prevented its being a Vertisol by defini- tion prior to the September 1980 National Soil Handbook Notice. The upper horizon of this series is usually a massive surface crust a few cm thick. The lower horizons crack deeply and have ma-skelsepic plasmic fabric (Fig. 6c). The lower horizons contain secondary gypsum (Fig. 6d), suggesting that this is the limit of turbation, or else that turbation takes place very slowly in this soil.

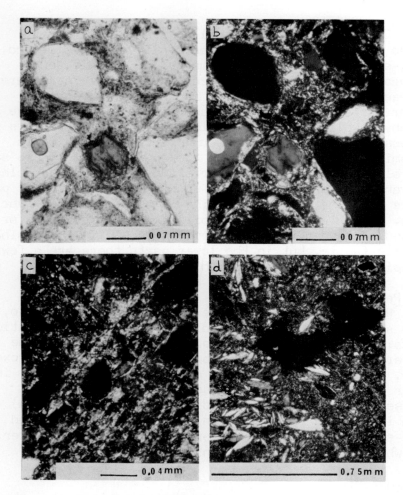

Fig. 6.

Turbation

Turbation in Vertisols and vertic subgroups probably
results from differential wetting of dry soil. Some model
experiments are helpful in showing how externally applied
stress can produce orientation similar to that found in
these soils. Maximum preferred orientations are found to
be perpendicular to the greatest shortening in deformed
clay materials (Clark, 1970). However, a large amount
of compaction is required to produce even a moderately

strong preferred orientation. In distortion experiments,
the strongest preferred orientations are developed at
lower water content. Substantially weaker preferred
orientations for the same amount of total strain were
produced in the more porous clay and those containing a
high electrolyte content (Clark, 1970). In kaolinite
samples with a porosity of 50% he found that preferred
orientation increased markedly at a strain of about 2.
He measured strain as $1 \div (1 - \%$ shortening). The
preferred orientation ratio was measured by X-ray diffraction.

Brewer (1964) suggests that vosepic plasmic fabric is
produced by forces just insufficient to cause shearing of
the soil matrix. Masepic plasmic fabric, he believes,
is produced by the shearing of clay-size soil material as
described by White and Bonestell (1960). Brewer considered omnisepic fabric as being produced under extreme
conditions of pressure. In the samples examined,
omnisepic plasmic fabric occurs in the Vertic Hapludalfs
but not in the Vertisols. Some shearing of clay-size
material has occurred in the Vertic Hapludalfs but they
lack intersecting slickensides. The linear extensibilities of the two groups of soils are about the same;
but either the vertic subgroups do not dry as often or as
thoroughly as the Vertisols, or else in the Vertic
subgroups, there is less movement of surface horizons

Fig. 6. (a,b) Porphyroskelic related distribution in
        the A11 horizon (10-46 cm) in Bosanko pedon 37-2.
        Note the incomplete mixing of organic matter and
        plasma. The grain in the centre of figure is
        hornblende, the other large grains are felds-
        pars. (a) PPL and (b) XPL; Note stress oriented
        clays around the grains; (c) The B23 horizon
        (20-36 cm) of Bickerdyke pedon 11-5 has ma-
        skelsepic plasmic fabric. This figure shows an
        area having weakly developed masepic plasmic
        fabric. XPL; (d) Gypsum filling voids and
        invading the matrix of the B34 horizon of Bicker-
        dyke pedon 11-4. The area is at the base of a
        crack at about 127 cm depth. XPL.

into the cracks.

McCormack and Wilding (1973) have proposed a mechanism for forming lattisepic fabric. They studied a Geeburg soil (fine illitic, mesic Aquic Hapludalf) which at saturation has a shear strength of 0.5 to 0.8 kg/cm^2. They found swelling pressures of 1 to 2 kg/cm^2 are generated as the soil-water content increases from tensions of about 15 atmospheres to saturation and linear extensibilities of 5 to 8% through the same moisture contents. They found lattisepic fabric in the upper part of the argillic horizon, and some masepic fabric in the lower part. During the wetting cycle they picture more movement of the soil in a horizontal direction, but later upward pressures, which exceed the weight of the overlying soil, produce vertical rise. They believe that this movement of the plasma produces the lattice-shaped orientation.

CONCLUSIONS

The soils in this study are clayey. They have deep, wide cracks that open to the soil surface or to a thin, massive surface horizon at some time during the year, and they have high bulk densities between the cracks. Some of the soils do not have slickensides or other direct evidence of turbation. There are no chemical and physical differences between the Vertisols and the Vertic Torrifluvents and B2t horizons of the Vertic Hapludalfs studied. There are, however, differences in their morphology and micromorphology. The plasmic fabrics of vertic subgroups show no strain, i.e., are silasepic for example, or they show more or less balanced three-dimensional strain as in omnisepic, lattisepic, or skelsepic plasmic fabrics. Plasmic fabrics of vertic subgroups are not expected to be masepic because soils with dominantly masepic plasmic fabric would be expected to have intersecting slickensides and, hence, would be Vertisols. Most of the Vertisols have skelsepic plasmic fabric in upper parts and masepic, or an intergrade toward that kind of fabric, in lower parts. Turbation likely results from differential wetting of dry soil. The soils wet deeply because of the cracks which are open to the surface. This causes the soil at the bottom of the

cracks to wet first. In many instances, the expansion
of this wet soil forces the column to rise.

REFEREMCES
Baldwin, M., Kellogg, C.E. and Thorp, J. 1938. Soil
     classification. In: Soils and Men. U.S.D.A. Yearbook,
     979-1001.
Bellinfante, N., Paneque, G., Olmedo, J. and Baños, C.
     1974. Micromorphological study of Vertisols in
     southern Spain. In: G.K. Rutherford (Ed), Soil
     Microscopy.   The Limestone Press, Kingston,
     Ontario, 296-305.
Brewer, R. 1964. Fabric and Mineral Analysis of Soils.
     John Wiley & Sons, New York. 470 pp.
Clark, B.R. 1970. Mechanical formation of preferred
     orientation in clays. Amer. J. Sci., 269, 250-266.
De Vos t. N.C., Virgo, J.H. and K.J. 1969. Soil
     structure in Vertisols of the Blue Nile clay plains,
     Sudan. J. Soil Sci., 20, 189-206.
Gallavan, R.C. and Greene-Kelly, R. 1974. The effect of
     the evaporation of entrained liquid on soil fabric.
     J. Soil Sci., 25, 498-504.
Greene-Kelly, R. and Mackney, D. 1970. Preferred orientation
     of clay in soils: the effect of drying and wetting.
     In: D.A. Osmond and P. Bullock (Ed), Micromorpho-
     logical Techniques and Applications. Soil Survey
     Tech. Monogr. 2. Harpenden, England.
Hole, F.D. 1961. A classification of pedoturbations and
     some other processes and factors of soil formation
     in relation to isotropism and anisotropism. Soil Sci.,
     91, 375-377.
Innes, R.P. and Pluth, D.J. 1970. Thin section preparation
     using an epoxy impregnation for petrographic and
     electron microprobe analysis. Soil Sci. Soc. Amer.
     Proc., 34, 483-485.
Jongerius, A. and Bonfils, C.G. 1964. Micromorphologia de
     un suelo negro grumosolico de la provinicia de Entre
     Rios. Investigationes Agropecuarias, Serie 3. 1,
     33-53.
Katti, R.K., Lal, R.K., Fotedar, S.K. and Kulkarni, S.K.
     1969. Depth effects in expansive clays. Proc. Int.
     Research and Eng. Conf., Expansive Clay Soils, 2nd
     Texas A&M. Univ. Press, College Station, Texas.

362-373.

Knight, M.J. 1980. Structural analysis and mechanical origins of gilgai at Boorook, Victoria, Australia. Geoderma, 23, 245-283.

Komornik, A. and Zeitlin, J.G. 1970. Laboratory determination of lateral and vertical stresses in compacted swelling clay. J. Materials, 5, 108-128.

Lafeber, D. 1965. The graphical representation of planar pore patterns in soils. Aust. J. Soil Res., 3, 143-164.

McCormack, D.E. and Wilding, L.P. 1973. Proposed origin of lattisepic fabric. In: G.K. Rutherford (Ed), Soil Microscopy. The Limestone Press, Kingston, Ontario, 761-771.

Nemeçek, J. 1975. Micromorphological marks of illuvial colloids accumulation in the soil profile (1st argilluviation). Vedecke Prace, Vyzkumnych Ustavu Rostlinne Vyroby Praze-Ruzyni, 20, 81-88.

Rode, A.A., Yarilova, Y.A. and Rashevskaya, V.V. 1960. Soils of large depressions. Soviet Soil Sci., 799-809.

Sleeman, J.R. 1963. Cracks, peds, and their surfaces in some soils of the Riverine Plain, New South Wales. Aust. J. Soil Res., 1, 91-102.

Soil Survey Staff. 1975. Soil Taxonomy: a basic system of soil classification for making and interpreting soil surveys. Agricultural Handbook 436. U.S.D.A. Washington, D.C. 754 pp.

Thorp, J. and Smith, G.D. 1949. Higher categories of soil classification: order, suborder, and great soil groups. Soil Sci., 67, 117-126.

Towner, G.D. 1961. Influence of soil-water suction on some mechanical properties of soils. J. Soil Sci., 12, 180-187.

U.S. Soil Conservation Service. 1972. Soil survey laboratory methods and procedures for collecting soil samples. Soil Survey Investigations Report No.1. 63 pp.

White, E.M. and Bonestell, R.G. 1960. Some gilgaied soils in South Dakota. Soil Sci. Soc. Amer. Proc., 24, 305-309.

MICROSCOPY OF THE CAMBIC HORIZON

P. Aurousseau

Laboratoire de Science du Sol,
65 rue de Saint-Brieuc,
35042 Rennes,
France

ABSTRACT
The cambic horizon as defined by the Soil Taxonomy
does not have specific microscopic features. A range of
kinds of weathering, soil structure and microstructure,
soil porosity, matrix and pedological features have been
described.

The genetic significance as expressed by the type
of weathering or by the pedological features varies with
the kind of soil, testifying to the large "variance" of
the concept of the cambic horizon.

Soil microscopy can help to distinguish and define
several kinds of cambic horizons, as subsets of low
variance, within the cambic concept.

INTRODUCTION
Among systems of soil classification prior to Soil
Taxonomy (Soil Survey Staff, 1975), concepts of
weathering B horizons were very broad.   In Soil
Taxonomy pedologists now have at their disposal horizon
concepts exclusive of each other. In this paper, the
main points of the definitions of cambic horizons are
first summarised, and the subject is then reviewed.
Finally the contribution of soil microscopy to the
advancement of knowledge of such horizons is examined.

SUMMARY OF THE CAMBIC HORIZON DEFINITION
The cambic horizon is an altered horizon. Physical
alteration is the result of movement of soil particles
to such an extent as to destroy most of the original rock
structure, or to aggregate the soil particles into peds,
or both. Chemical alteration is the result of hydrolysis,
solution redistribution, removal of carbonates, iron
oxides or other soluble compounds.

The cambic horizon is immediately below one of the
diagnostic epipedons (mollic, anthropic, umbric, histic,
plaggen ochric). It normally lies in the position of a B
horizon. The concept is confined to an horizon which
lacks features associated with mineral accumulation or
extreme weathering. For this reason an horizon transitional to another with more strongly expressed genetic
features is excluded. Below many argillic and spodic
horizons, there is an horizon transitional to the C
horizon, with comparable characteristics to those of a
cambic horizon. Such an horizon is not considered cambic
because of its position in the profile. Transitional
horizons such as A3 or B1 have properties of a cambic
horizon. They fit the definition if they are clearly
not transitional to an argillic horizon. Thus, position
as well as alteration without significant illuviation
are important characteristics of the cambic horizon.
The genetic significance of the cambic horizon varies
somewhat with the kind of soil. Several kinds of cambic
horizon can be defined but the limits of the transitional
forms would be difficult to recognise.
Soil Taxonomy distinguishes several kinds of cambic
horizon:
  - formed in the presence of fluctuating groundwater-
    table.
  - formed in the absence of both groundwater and
    carbonates.
  - formed from highly calcareous material.
A few points in this definition need emphasis:

1. Position criteria are important in the definition of
    the cambic horizon. Transitional horizons that have
    the same intrinsic characteristics as cambic horizons
    may not be considered cambic because of their position.
2. Very few characteristics are specific to the cambic
    horizon and the concept has a very large variance.
3. The genetic significance varies with the kind of soil.
4. Genetic features are weakly expressed, hence the
    difficulty in studying such kinds of horizon by micro-
    scopic methods which in most cases would be suitable
    for detailed study of genetic features and processes
    in soil horizons. The bibliography of cambic horizons
    is, as a result, rather poor.

BIBLIOGRAPHIC REVIEW
        The bibliographic review can be considered under
four headings;
        Weathering
        Structure - Porosity
        Matrix
        Pedological features

Weathering
        There are a number of difficulties associated with
the recognition of weathering in cambic horizons. Primary
minerals are often smaller in cambic horizons than in
associated C horizons. It is difficult to determine the
optical properties, e.g. interference figures in
convergent light, of small minerals. B horizons are
often subject to pedoturbation, integrating secondary
weathering products in the matrix.
        In spite of the fact that weathering is one of the
main processes characteristic of cambic horizons, there
is very little information on the subject because of the
limiting magnification of the optical microscope and
because of the difficulties of linking primary minerals
and secondary products in the same sample.
        Nevertheless, because of the wide range of pedological
conditions that can occur in cambic horizons, a subdivision could be made on the nature of the weathering
environment: acid, hydrolytic, alkaline, hydromorphic
and andic.

Structure - Porosity
        According to the definition of the cambic horizon,
structure is one of the main criteria by which such
horizons are distinguished from other diagnostic horizons.
However, numerous kinds of structure have been described
in cambic horizons:
1. Weak prismatic and moderate blocky in calcareous cambic
    horizons (Gile, 1966).
2. Strongly developed prismatic in calcareous vertic
    horizons (Bellinfante et al., 1974).
3. Blocky structure defined by interconnected vughs,
    channels and planes in Andosols (Benayas et al., 1974).

4. Rounded micropeds with strongly interconnected vughs
    in Andosols (Benayas et al., 1974).
5. Soil matrix organised into rounded aggregates with
    a mean diam. of 100 μm in red soils (Fedoroff and
    Rodriguez, 1978).
6. Polyhedric aggregates with polyconcave voids in
    Andosols (Kawai, 1969; Bech et al., 1977).
7. Free packing and close packing of rounded aggregates
    of 70 - 100 μm; incompletely defined polyhedric
    aggregates with polyconcave voids in acid brown
    soils (Aurousseau, 1978).
8. Massive structure,and blocky structure incompletely
    defined by vughs,in eutric brown soils (Rudeforth,
    1966, 1967; Clayden 1970).
    Clearly, there is no structure specific to the
cambic horizon. Several kinds of cambic horizon can be
defined according to their soil structure and microstructure. There is also a rather close relationship
between structure and microstructure and the pedological
conditions.

Matrix
    Two main features have been described in the matrix:
the plasmic fabric and the plasma. The definitions of
plasmic fabric proposed by Brewer (1964) have been
extensively used in soil descriptions. Many types of
plasmic and elementary fabric have been recognised in
cambic horizons: skelsepic, insepic, vosepic,
clinosepic, silasepic, intertextic, porphyroskelic,
agglomeroplasmic. No particular plasmic fabric
appears to be specific to cambic horizons. Whether
there is a relationship between plasmic fabric described
in cambic horizons and other microscopic parameters of
the cambic horizon has never been discussed and only
a few papers give information about the internal nature
of the matrix:
1. A dusty yellowish brown and reddish brown matrix
    with micro-inclusions is described in Andepts
    (Bech et al., 1977).
2. A light coloured matrix, yellow to yellowish brown,
    is described in Ochrepts (Fedoroff and Aurousseau,
    1981).

3. A poorly separated plasma, composed mainly of iso-
    tropic to very weakly birefringent material,is
    described in Andepts (Kawai, 1969).

Pedological Features
    By definition, pedological features are weakly
expressed in cambic horizons. The following descriptions
have been noted in the literature:
1. Carbonate accumulation in the form of filaments on
    ped surfaces and along root channels.
2. Distinct oriented clay coatings on sand grains and
    pebbles in a calcareous cambic horizon (Gile, 1966).
3. Infrequent argillans surrounding peds in Andepts
    (Aguilar and Delgado, 1974).
4. Black Mn-Fe diffuse nodules in vertic calcareous
    soils.
5. Dynamic horizons with papules in red Mediterranean
    soils (Fedoroff and Rodriguez, 1978).
6. Grain edges with sharp boundaries (Gile, 1966;
    Fedoroff and Aurousseau, 1981).
    Depending on pedological conditions the cambic
horizon may be affected by translocation of carbonates,
clay, iron. Sometimes it represents a first step towards
more strongly expressed horizons with greater genetic
significance, such as albic or argillic. Sometimes the
cambic horizon is in equilibrium with pedological
conditions which produce only weak soil development.

CONTRIBUTION OF SOIL MICROSCOPY TO IMPROVING KNOWLEDGE
OF CAMBIC HORIZONS
    The cambic horizon does not have specific microscopic
features. Different kinds of weathering, soil structure,
soil porosity, matrix and pedological features have
been described. Clearly, the genetic significance as
expressed by the type of weathering or by the pedological
features varies with the kind of soil, testifying to
the large variance of the concept of cambic horizon.
    Soil microscopy can be used to distinguish and define
several kinds of cambic horizons. To proceed, microscopic
fabrics and horizons of low variance, subsets of the
concept of cambic horizon, need to be described and defined.
Sufficent data are available to define some such subsets
of the cambic horizon.

For example: In andic conditions where weathering is
         specific, a principal type of structure and porosity
         has been described by Kawai (1969), Bech et al.
         (1977), Fedoroff and Rodriguez (1978):
Small brown aggregates 30 - 40 μm diam. with rounded or
with very smooth edges are randomly distributed in the
horizon. Generally they are composed of fragments of
brown organic matter, black or reddish black iron nodules
and fragments of primary minerals randomly distributed
in a fine mass.
     Such a microscopic fabric defines an andic subset.
Soil scientists specialising in microscopy should produce
more information to define further subsets of cambic
horizons.

BIBLIOGRAPHY

Aguilar, J. and Delgado, M. 1974. Micromorphological
     study of soils developed on andesitic rocks in
     oriental Andalusia (Spain). In : G.K. Rutherford
     (Ed), Soil Microscopy. The Limestone Press,
     Kingston, Ontario, 281-295.
Aurousseau, P. 1978. Caractérisation micromorphologique
     de l'agrégation et des transferts de particules
     dans les sols bruns acides. Cas des sols sur
     granite du Morvan (France). In : M. Delgado (Ed),
     Micromorfologia de Suelos. Universidad de Granada,
     655-667.
Bech, J., Fedoroff, N. and Sole, A., 1977. Etude des
     andosols d'Olot (Gerona, Espagne). Cah. ORSTOM,
     ser. Pédologie, 15, 4, 381-390.
Bellinfante, N., Paneque, G., Olmedo, J. and Baños, C.
     1974. Micromorphological study of vertisols in
     southern Spain. In : G.K. Rutherford (Ed), Soil
     Microscopy. The Limestone Press, Kingston, Ontario,
     296-306.
Benayas, J., Alonso, J. and Fernandez Caldes, E. 1974.
     Effect of the ecological environment on the micro-
     morphology and mineralogy of andosols (Tenerife,
     Island). In : G.K. Rutherford (Ed), Soil
     Microscopy. The Limestone Press, Kingston, Ontario,
     306-320.

Brewer, R. 1964. Fabric and Mineral Analysis of Soils.
    John Wiley and Sons, New York, 470 pp.
Clayden, B. 1970. The micromorphology of ochreous B
    horizons of sesquioxidic brown earths developed in
    upland Britain. Soil Survey Tech. Monogr. 2,
    Harpenden, England, 53-67.
Fedoroff, N. and Rodriguez, A. 1978. Comparaison
    micromorphologique des sols rouges des iles Canaries
    et du bassin mediterraneen. In : M. Delgado (Ed),
    Micromorfologia de Suelos. Universidad de Granada,
    867-929.
Fedoroff, N. and Aurousseau, P. 1981. Micromorphologie
    des sols bruns acides sur matériaux granitiques.
    Can. J. Soil Sci., 61, 483-496.
Gile, L.H. 1966. Cambic and certain non cambic horizons
    in desert soils of southern New Mexico. Soil Sci.
    Soc. Amer. Proc., 30, 773-781.
Kawai, K. 1969. Micromorphological studies of andosols
    in Japan. Bull. Nat. Inst. Agric. Japan, 145-154.
Rudeforth, C.C. 1966. The nature, distribution and
    origin of the soils of mid-Wales. M.Sc. Thesis,
    London Univ.
Rudeforth, C.C. 1967. Upland soils from Lower Palaeozoic
    sedimentary rocks in mid-Wales. Rep. Welsh Soils
    Disc. Gp., 8, 42-51.
Soil Survey Staff, 1975. Soil Taxonomy. A basic system
    of soil classification for making and interpreting
    soil surveys. Agricultural Handbook 436, U.S.D.A.,
    Washington, D.C.

MICROMORPHOLOGICAL ADVANCES IN ROCK WEATHERING STUDIES

A. Meunier

Laboratoire de Pédologie de L'Université de Poitiers,
E.R.A. 070.220 du C.N.R.S., "Pédologie des Pays Atlantiques"
40, avenue du Recteur Pineau, 86022 POITIERS, France

ABSTRACT
    The first studies on weathered rocks used micro-
morphological observations only as illustrations.
Nowadays, these observations are the fundamental support
of any work on natural or artificial weathering. Rocks
may be considered as discontinuous materials composed
of several crystalline species and of a few empty spaces.
These different components were studied separately.
    The de-stabilisation of each primary mineral con-
sidered as isolated systems has been investigated.
High-resolution microscopy and selected area diffraction
studies supplement usefully micromorphological and
chemical data. They allow accurate descriptions of
recrystallisation processes in the weathered crystals.
The studies of microcrack features and distribution by
colouration techniques or SEM observations define the
earlier fluid paths in the rocks. They explain the
heterogeneous behaviour of the crystalline material
during alteration.
    Petrographical studies of the weathered rocks
supported by such data emphasise the importance of the
microsystem concept. By this means descriptive micromorphology becomes an interpretative discipline.

INTRODUCTION
    Rock weathering has been known as a geological
process for a long time. Exchanges between minerals and
the weathering solution were discovered very early. In
France, Brongniart (1807) observed that feldspar
kaolinisation depends on groundwater action. Ebelmen
(1847) published the first geochemical data from
weathered basaltic rocks. The earliest micromorphological
descriptions of weathered minerals appeared 50 years
later with the works of Lacroix (1896) and de Lapparent

(1909). Micromorphology has really existed as an
independent discipline since the 1960s. Its development
depended on technical advances in making thin sections
from incoherent rocks and especially in the routine use
of synthetic resin impregnation.

Rock weathering studies include three complementary
investigations : observations of natural weathered
materials, experimentation with artificial weathering,
and theoretical simulations. Micromorphological data
were used first as illustrations of geochemical studies
in natural or artificial occurrences. They have never
been used as supports for thermodynamical investigations.
This situation has changed over the last few years. Now
micromorphology has become the fundamental support of
any work on natural or artifical weathering phenomena.

Papers with micromorphological descriptions are
too numerous to be all cited here. Only some specific
modern aspects with a few chosen samples will be given.
The first geochemical investigations defined the general
processes of weathering but failed to explain the
variable behaviour of crystalline rocks. These are
composed of various crystal species with a few voids
which are the first paths of the weathering solution.
Therefore,the studies of minerals considered as isolated systems and the descriptions of rock microcracks
will be summarised. These specific data are of a great
interest for petrographic investigations of weathered
rocks and have helped to develop the microsystem concept.

GENERAL GEOCHEMICAL STUDIES
A rock is a heterogeneous material of different
species of minerals and few voids occurring as fine
cracks between minerals. In weathering conditions, water
flows into these microcracks and reacts with the surfaces
of the minerals. Incongruent dissolution of the various
primary silicates gives less dense secondary silicates
plus many more voids. The problem is to determine the
mechanism of mineral dissolution.

The first systematic investigations on weathered
rocks aimed at establishing global geochemical
descriptions of profiles developed in various conditions
of parent rock and climate (Harrison, 1933). Micromorphology was little used and only to illustrate some

specific phenomenon. In most cases, geochemical data
were associated with secondary mineral sequences,
established level by level in the profile (Bonifas,
1959; Grant, 1963; Millot, 1964; Wolff, 1967;
Tardy, 1969). Micromorphological investigations
were not used in experimental work on artificial
weathering (Pedro, 1964; Trichet, 1970). Theoretical
studies tried to simulate the global behaviour of rocks
(Garrels and Christ, 1959; Helgeson, 1968).
      The results obtained by experiments or theory
associated with observations of natural occurrences are
useful in describing the migration of elements on a
country or continental scale. They are not able to
explain the reality of weathering mechanisms which give
rise to these migrations. Crystalline rocks behave in
many different ways during alteration. The observation
of a thin section in granite saprolite shows the great
complexity of the processes. The primary texture of the
rocks is not modified to any extent but some parts are
replaced by microcrystalline porous plasma (*) in which
new silicate species appear. In other parts the original
minerals are still recognisable but they are locally
altered along internal microcracks or along intergranular
joints. These reactions produce phyllosilicate phases
different from those of the plasma. Thus, it is easy
to see that different secondary mineralogical facies
can coexist in very small volumes of rock (thin section
scale). Active microsystems are scattered and their
geochemical characteristics strictly depend on local
conditions. In such a way, micromorphology becomes the
necessary tool for petrographic investigations.

(*)   Plasma is composed of primary mineral debris
       surrounded by secondary minerals (clays, oxides...)
       and is characterised by a high porosity. Secondary
       minerals are often, but not always, of sub-
       microscopic dimensions. In some cases, neogenetic
       crystals are 20-50 μm long.

PRIMARY MINERALS CONSIDERED AS ISOLATED SYSTEMS
     Most sedimentary or crystalline rocks have a major
mineral whose alteration has most influence on the
development of the profile. Numerous papers have
described the secondary product sequences developed from
biotites, pyroxenes, amphiboles, plagioclases,
glauconites... (e.g. Delvigne, 1965; Bisdom, 1967;
Wilson and Farmer, 1970; Basham, 1974; Courbe et al.,
1981). In a few cases, artificial weathering experiments
and theoretical simulations reproduce observations of
natural phenomena and allow more accurate interpretation.
They sometimes reveal the existence of an unobserved
naturally occurring phase.
     A good example of such a problem is found in
orthoclase alteration studies. Its macroscopical
weathering features have been observed for a long time
in weathered granite profiles. Micromorphological studies
have shown that weathering develops an intense micro-
fracturing associated with the formation of secondary
crystals. The nature of these phases depends on stable
conditions. They are generally associated in multimineral
paragenesis (Bisdom, 1967; Seddoh, 1973; Meunier, 1980).
The exchange intensity between feldspars and solutions
has been measured experimentally. Using these data
some authors suggested destabilisation -recrystallisation
patterns for weathering orthoclase (Garrels and Howard,
1959; Lagache, 1965; Wollast, 1967). To explain
the decreasing exchange between fluid and feldspar with
time, Wollast (1967) suggests that a protective residual
layer forms at the interface.
     Such an amorphous coating, which is well developed
in weathered plagioclase (Delvigne and Martin, 1970),
has never been observed in orthoclase. Using very
high resolution electron microscopy, Eggleton and Buseck
(1980) have shown that natural alteration of this
feldspar produces negative shaped pits in which an
amorphous phase precipitates. In a few cases, they were
able to observe the progressive organization of 10 $\overset{o}{A}$
layers within the amorphous phase. In this case, the
ultramicromorphological investigation gives an answer
to the experimental hypothesis and clearly shows that

secondary phyllosilicate phases in a natural weathered
orthoclase have an amorphous precursor.

Another example of progress in micromorphology by
using an ultramicroscopical technique is found in the
studies of biotite weathering.   These micas are the
most reactive minerals in weathering conditions and
produce various secondary phases. Numerous papers
have been published on the subject (Coleman et al.,
1963; Newman and Brown, 1966; Wilson, 1966; Ismail,
1970; Meunier and Velde,1978). The exchange processes
of biotites with solutions were investigated by Robert
(1971) who showed the great importance of pH in the
lattice destruction. Hoda and Hood (1972) have
demonstrated the influence of Ca, Mg, Na and K
activities in the crystallisation of secondary phases.
These experimental studies are very helpful for the
geochemical interpretation of observations of natural
weathered biotites. However, the main mechanisms
causing the destruction of the micaceous lattice and
of secondary phases crystallisation still remain
unknown.

By using selected area diffraction techniques
(SAD), Gilkes and Suddhiprakarn (1979a,b) have begun
to solve this problem. They first studied naturally
weathered biotites by classical methods : XRD,
polarising microscope observation and electron microprobe. With these techniques, secondary minerals were
identified, then their organisation within the original
crystal was investigated by SAD. The authors have
demonstrated the strong influence of the biotite
lattice on the growth of kaolinite, vermiculite, mixed-
layer minerals, gibbsite and goethite. In this case,
SAD may be considered as a direct morphological
technique that gives reciprocal lattice pictures. Such
information is helpful in explaining a very common
phenomenon in biotite weathering : pseudomorphism.

From a methodological point of view these two
examples demonstrate the necessity of using quantitative
techniques in association with classical micromorphological
ones. Optical determinations, XRD identifications and
electron microprobe data remain the basic methods for
petrographical studies. They are sometimes fruitfully

completed by electron microscopical techniques. In
this way the descriptive micromorphology becomes
interpretative. In 1973, Brewer postulated :
"Micromorphology is a discipline at the chemistry-
mineralogy interface".

ROCK MICROCRACKS
     Porous and fissure systems of weathered rocks and
soils have been described for a long time. They have
been classified on their macro- or micro-organisation
and on their mineral or organic filling by Brewer (1964).
In contrast, the primary porosity of fresh rocks is not
so well known. Direct observations of submicroscopic
voids is not very easy. Such an investigation necessitates
colouration techniques (Garrels et al., 1949; Perami,
1971) or special preparations for SEM studies (Simmons
and Richter, 1976).
     Crystalline rocks contain three types of empty
spaces : intergranular joins, intramineral microcracks,
and multimineral microcracks. In most cases, they may
be compared to irregular discontinuous surfaces. Sometimes they are tube-shaped, especially in micas. Their
size varies between 0.1 and 0.01 μm (Richter and Simmons,
1977). They are the consequence of mechanical or thermal
stresses.
     Microcracks constitute the first paths for the
weathering solution into the rock. Dissolution features
and secondary phase growths occur in these porous spaces.
In the earliest stage of weathering of a granite, an
illitic mica crystallisation is observable in K-feld-
spars-muscovite contacts (Meunier and Velde, 1976). The
evidence suggests that these reactions depend strictly
on the microcrack distribution in the rock. An investigation of voids in an unweathered granite by methylene blue
colouration (10 bars pressure injection over 143 hours)
statistically shows that intergranular contact microcracks are more frequent than intra or multimineral
ones. They appear along quartz-feldspar (63%) and
quartz-mica contacts (13%), but are very rare in feldspar-
feldspar contacts. These data explain the varying
behaviour of crystalline rocks in weathering conditions :
some local sites are active and transform faster than
others.

Progress in the knowledge of weathering mechanisms
(destabilisation of primary minerals and recrystallisation
of a part of the material produced) in these early stages
will depend on investigations of the physicochemical
properties of aqueous fluids in such tiny spaces.
Capillary forces and surface effects strongly influence
exchange reactions between minerals and solutions. In
this way, some experimental studies have tried to give
quantitative information on rock-fluid exchanges and
migration of elements (Thenoz, 1966; Baudracco, 1978).
The authors aimed to define a weathering coefficient for
the rock studied.

CHARACTERISATION OF ACTIVE CHEMICAL MICROSYSTEMS
Petrographic investigations of various weathered
rocks have shown that reactions occur in small-scale
sites where micro-environmental conditions are important
(Proust and Velde, 1978; Meunier and Velde, 1976, 1978;
Ildefonse, 1980). Reaction sites are identified by
optical observations at each level of the profile.
Then, secondary phases are isolated by several methods :
 (a) Ultrasonic treatments on separated primary minerals,
 (b) Direct sampling by micromanipulations,
 (c) Concentration by density or magnetic techniques.
These phases are identified by XRD and infrared
absorption. Returning to microscopical observations,
it is then possible to determine the distribution of the
identified phases in some chosen sites by microprobe
analysis. An energy dispersive system (Tracor Northern
Instrument) is used because the electron beam energy
is low (15 Kv, 1 nanoampere). In this way phyllosilicates
are not destroyed before the end of the count-time (120 s).
The ZAF correction program, calibration and choice of
standards are adapted to the phyllosilicate chemistry
problem (B. Velde, pers. comm.).
This step method gives reasonable accuracy for the
crystallo-chemical determinations of secondary phases
in their reaction micro-site. A petrographic interpretation of the micromorphological features described
earlier then becomes possible. We have defined three
types of micro-systems in the weathering processes:
contact, plasmic and fissural.   These microsystems can
coexist inside a single thin section because of rock

heterogeneity. The macroscopic nature of the material
depends on the predominance of one of these microsystems.

Contact Microsystems
     One finds that the first reaction zone in the altera-
tion process is between grains of different mineral species.
Obviously the water circulates slowly and will contain a
relatively high concentration of dissolved material which
reflects the chemical character of the adjacent minerals.
In granites, white mica is the first phase to form in zones
where potassium feldspar joins magmatic micas (Fig. 1a).
The new phases have a distinct composition compared to the
phases formed at higher temperatures (Meunier and Velde,
1976). In a metagabbro, plagioclase is weathered at its
contact with amphibole where a ferric beidellite is
formed (Ildefonse, 1980). The beidellites contain more
alumina the further they are from the amphibole (i.e.
further within the feldspar) (Fig. 1b).

Plasmic Microsystems
     Two types of plasmic systems were identified which
often occur within the same thin section. The primary
plasma manifests itself when the primary minerals
become unstable in themselves. It is made up of small
mineral debris of the host mineral associated with new
clay minerals which are often multiphase. At times a
distinct chemical zonation of the new minerals can be
seen (Proust and Velde, 1978; Fig. 1c). The primary
plasmic system develops mainly in horizons where the rock
conserves its petrographic structure and fabric. It
can continue however in isolated grains within horizons
which have lost the initial rock structure.
     Secondary plasma is found in horizons where the
original grain boundaries of the rocks are lost due to
intense argillisation. Transport of clay is typical in
this horizon. The parent mineral debris is isolated
in a clay-pore matrix which effects a mixing of the clay
phases of recent crystallisation : this creates new
chemical systems, which include all of the altered and
altering minerals. A pre-existing phase can change
composition, for example, ferric beidellites in gabbro.
Further, new phases are frequently produced. A well-
crystallised trioctahedral vermiculite (Fig. 1d) has

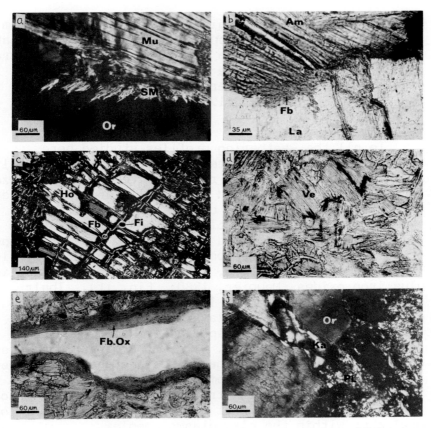

Fig. 1. (a) Secondary mica crystallisation (SM) at
    muscovite-orthoclase (Mu:Or) contact (Granites of
    Massif Armoricain);   (b) Ferriferous beidellite
    crystallisation (Fb) at amphibole - labradorite
    (Am:La) contact (Gabbro of Massif du Pallet); (c)
    Internal destabilisation of hornblendes (Ho).
    Extinct zones are ferriferous intergrades (Fi),
    bright zones are ferriferous beidellites (Fb)
    (Amphibolites of Limousin);   (d) Secondary vermicu-
    lite crystallisation (Ve) in restructured zones of
    weathered gabbro (Massif du Pallet);   (e) Ferri-
    ferous beidellite and oxide deposits (Fb-Ox) on
    edges of a crack (Gabbro du Pallet);   (f) Kaoli-
    nite (Ka) deposit in orthoclase (Or) crack (PL=
    plasma) (Granite of Massif Armoricain).

been identified in the argillised horizon formed from
a mica-free gabbro. Microprobe analyses indicate that
the chemistry of the phase is rather variable from grain
to grain but their average composition nevertheless
distinguishes them from vermiculites derived from
micas (Ildefonse et al., 1979).

Fissural Microsystems
     In all the weathering profiles studied here,
fissural systems were quite similar. They affect all
of the horizons where the rock is altered. Their size
varies from   μm to cm and their walls are often covered
with a layer of deposited clay. The composition of
these coatings is remarkably constant throughout the
profiles. In amphibolites and gabbros, they are composed
of ferric beidellites associated with iron oxides (Fig.
le). In altered granites, they are composed of kaolinite
plus iron oxides (Meunier, 1980; Fig. 1f). Their
composition is constant. This indicates that transported and re-deposited clay will control the
composition of dissolved species in solutions which
pass through the fissures. Since the mineralogy is so
constant in the profile one can deduce that these are not
active clay-forming systems but simply the material
transferred. The active transformation of rocks into soil
occurs within the rocks at grain boundaries, within old
grains or within the clay plasma complex.
     In temperate climates, the fissural system is
relatively reduced compared to the volume of altering
rock,while in tropical climates the mineralogy of the
fissures becomes dominant and the initial stages of
alteration much more reduced.
     Accumulation layers may be very thick and sometimes
form important concentrations of Al, Mn, Fe or Si. These
crusts behave as new parent rocks and are weathered in
different ways according to stationary conditions (climate,
topography, leaching intensity...). Dissolution and
migration of elements induce new structural distribution
of gibbsite, iron hydroxides, siliceous phases (Nahon,
1976; Bocquier, 1976; Fouillac et al., 1977).
     Some sedimentary rocks present the same weathering
patterns as crystalline ones. Glauconitic sands in
which the grains are of macro- or microscopic scale

produce differentiated profiles (Courbe et al., 1981;
Loveland, 1981). The nature of secondary phases
produced from glauconite depends on the microsystem
conditions : Al-Fe mixed layers appear first in
glauconitic plasmic microsystems, then nontronite-
kaolinite associations crystallise in argillaceous plasmic
microsystems and, finally, kaolinite and iron oxides
in fissural microsystems.
     Data on separated minerals considered as isolated
systems and rock porosity observations may be fruitfully
used in petrographical investigations of weathered rocks.
Local small-scale reactions depend on mineral-microvugh
relations. By the means of the microsystem concept one
can explain the presence of different alteration facies
inside a single thin section. In addition, it constitutes
a method to distinguish deuteric or hydrothermal
paragenesis from that of weathering. This was demonstrated in the study of micaceous phases in a granite
saprolite (Meunier and Velde, 1976).   As a consequence, experimental investigations or theoretical
simulations considering local equilibria may be more
accurate than general ones. Korzhinskii (1959) described
an altered rock as a mosaic of independent geochemical
systems. Michard and Fouillac (1974) tried to apply
this concept in a 'boxes' theory of simulation. Each
system is compared to a box characterised by an opening
parameter which is defined by the ratio between
dissolution rate of the mineral and solution flow rate
inside the box.

CONCLUDING REMARKS
     Rock weathering studies show the same historic
evolution as classical petrography. Field observations,
which were the only possible contributions from the
earliest workers, were further developed by general
geochemical studies and now by accurate crystallochemical
investigations. The weathered material must be defined
at all scales of observation, from profiles to micro-
sites. Micromorphology is the necessary support of the
geochemical and crystallographical approach to the
weathering phenomena.
     This discipline will progress now in two different

A. Meunier

but complementary ways : identification of real
destabilisation-recrystallisation mechanisms and
quantification of weathered features. They both depend
on technical and methodological progress. The identification of secondary phases in a thin section is the
main problem for petrologists. Some authors have tried
to obtain X-ray diffraction patterns from selected areas
in thin sections (Wicks and Zussman, 1975; Wilson and
Clark, 1978). Others have tried to extract very small
quantities of matter from thin-sections with microtools
and micromanipulators (Wallace, 1955; Rickwood, 1977).
The results obtained were sometimes useful when they
concerned specific material but these methods have no
actual routine application. The association of X-ray
identification with microprobe data on the reaction
site may be better adapted to the petrology of
weathered rock. Progress must be realized on microsampling
of the material.
      In contrast, quantification of features is nowadays
a routine technique by the use of image analysis
computer systems (Bullock and Murphy, 1980). The
accuracy of data depends on the identification and
classification of significant weathering features
only. Stoops et al. (1979) have proposed a general
pattern for quantification studies. Such a pattern,
very useful for soils, will be of great interest for
altered rocks if the microsystem concept is used to
define the quantifiable features.

REFERENCES

Basham, I.R. 1974. Mineralogical changes associated
      with deep weathering of gabbro in Aberdeenshire.
      Clay Min., 10, 189-202.
Baudracco, J. 1978. Contribution a l'étude de l'altér-
      abilité des roches sous l'action des eaux
      naturelles. Thèse, Fac. Sci. Toulouse, 1 vol.,
      241 pp.
Bisdom, E.B.A. 1967. Micromorphology of a weathered
      granite near the Ria de Arosa (NW Spain). Leidse
      Geol. Meded., 37, 33-67.
Bocquier, G. 1976. Synthèse et perspectives : migrations
      et accumulations de l'aluminium et du fer. Bull.

de la Soc. Geol. de France, 7, 18, 69-74.

Bonifas, M. 1959. Contribution a l'étude géochimique de
    l'altération latéritique. Mém. du Service de la
    Carte Géol. d'Alsace-Lorraine, 17, 159 pp.

Brewer, R. 1964. Fabric and Mineral Analysis of Soils.
    John Wiley and Sons, New York, 470 pp.

Brewer, R. 1973. Micromorphology. A discipline at
    the chemistry-mineralogy interface. Soil Sci.,115,
    261-267.

Brongniart, A. 1807. Traité élementaire de minéralogie
    avec des applications aux arts. Déterville,
    Paris, 2 tomes.

Bullock, P. and Murphy, C.P. 1980. Towards the quanti-
    fication of soil microstructure. J. Microscopy,
    120, 317-328.

Coleman, N.T., Le Roux, F.M. and Cady, I.G. 1963.
    Biotite-Hydrobiotite-Vermiculite in soils. Nature,
    198, 409-410.

Courbe, Ch., Velde, B. and Meunier, A. 1981. Weathering
    of glauconites-reversal of the glauconitisation
    process in a soil profile in Western France.
    Clay Min. (in press).

Delvigne, J. 1965. La formation des minéraux secondaires
    en milieu ferralitique. Mémoires O.R.S.T.O.M., 13,
    177 pp.

Delvigne, J. and Martin, H. 1970. Analyse à la microsonde
    électronique de l'altération d'un plagioclase en
    kaolinite par l'intermédiaire d'une phase amorphe.
    Cahiers O.R.S.T.O.M., série Géologie, 2, 259-295.

Ebelmen, H. 1847. Recherches sur la décomposition des
    roches. Anns. des Mines, 12, 4-5-6, 627-654.

Eggleton, R.A. and Buseck, P.P. 1980. High resolution
    electron microscopy of feldspar weathering.
    Clays Clay Min., 28, 173- 178.

Fouillac, C., Michard, G. and Bocquier, G. 1977. Une
    méthode de simulation de l'évolution des profils
    d'altération. Geochimica Cosmochimica Acta, 40,
    207-213.

Garrels, R.M., Dreyer, R.M. and Howland, A.L. 1949.
    Diffusion of ions through intergranular spaces in
    water-saturated rocks. Geol. Soc. Amer. Bull.,
    60, 1809.

Garrels, R.M. and Christ, C.L. 1959. Solutions, Minerals

and Equilibria.   Harper and Row, New York, 450 pp.
Garrels, R.M. and Howard, P. 1959.   Reaction of feldspar
    and mica with water at low temperature and pressure.
    Clays Clay Min., Proc. 6th National Conf., 68-88.
Gilkes, R.J. and Suddhiprakarn, A. 1979a.   Biotite
    alteration in deeply weathered granite.   I. Micro-
    morphological, mineralogical and chemical pro-
    perties.   Clays Clay Min., 27, 349-360.
Gilkes, R.J. and Suddhiprakarn, A. 1979b.   Biotite
    alteration in deeply weathered granite. II. The
    oriented growth of secondary minerals.   Clays
    Clay Min., 27, 361-367.
Grant, W.H. 1963.   Weathering of Stone Mountain
    granite.   Clays Clay Min., Proc. 11th National
    Conf., 65-73.
Harrison, J.H. 1933.   The katametamorphism of igneous
    rocks under humid and tropical contiions.   Imp.
    Bur. Soil Sci. Rothamsted Exp. Sta., Harpenden,
    79 pp.
Helgeson, H.C. 1968.   Evaluation of irreversible reactions
    in geochemical processes involving minerals and
    aqueous solution.   I. Thermodynamical relations.
    Geochimica and Cosmochimica Acta, 37, 455-480.
Hoda, S.N. and Hood, W.C. 1972.   Laboratory alteration
    of triotahedral micas.   Clays Clay Min., 20,
    343-358.
Ildefonse, Ph. 1980.   Mineral facies developed by
    weathering of a meta-gabbro, Loire-Atlantique
    (France).   Geoderma, 24, 257-273.
Ildefonse,Ph., Copin, E. and Velde, B. 1979.   A soil
    vermiculite formed from a meta-gabbro, Loire-
    Atlantique, France.   Clay Min., 14, 201-210.
Ismail, F.T. 1970.   Biotite weathering and clay formation
    in acid and humid regions, California.   Soil Sci.,
    109, 257-261.
Korzhinskii, D.S. 1959.   Physico-chemical basis of the
    analysis of the paragenesis of minerals.   Translation
    of the Consultant Bureau, Inc. New York, 1 vol.,
    142 pp.
Lacroix, A. 1896.   Minéralogie de la France et de ses
    anciens territoires d'Outre-Mer.   Librairie

Scientifique et Technique, Paris. Rééd. 1962-1964, 6 tomes.

Lagache, M. 1965. Contribution a l'étude de l'altération des feldspaths dans l'eau, entre 100 et 200ºC, sous diverses pressions de CO_2 et application à la synthèse des mineraux argileux. Bull. de la Soc. Française de Minéral. et Cristallog., 88, 223-253.

Lapparent, J. de, 1909. Etude comparative de quelques porphyroides françaises. Bull. de la Soc. Française de Minéral., 32, 174-304.

Loveland, P.J. 1981. Weathering of a soil glauconite in Southern England. Geoderma, 25, 35-54.

Meunier, A. 1980. Les mécanismes de l'altération des granites et le rôle des microsystèmes. Etude des arènes du massif granitique de Parthenay (Deux-Sèvres). Thèse Faculté des Sciences de Poitiers. Mémoires de la Soc. Géol. de France, 140, 80 pp.

Meunier, A. and Velde, B. 1976. Mineral reactions at grain contacts in early stages of granite weathering. Clay Min., 11, 235-240.

Meunier, A. and Velde, B. 1978. Biotite weathering in granites of Western France. VI Int. Clay Conf., Oxford, 405-413.

Michard, G. and Fouillac, C. 1974. Evaluation des transferts d'éléments au cours des processus d'altération des minéraux par des fluides. Comptes Rendus de l'Acad. des Sci., Paris, 278 D, 2727-2729.

Millot, G. 1964. Géologie des Argiles. Masson, Paris, 1 vol., 499 pp.

Nahon, D. 1976. Cuirasses ferrugineuses et encroûtements calcaires au Sénégal occidental et en Mauritanie. Systèmes évolutifs : géochimie, structures, relais et coexistences. Thèse, Faculté des Sciences de Marseille. Mémoires Sci. Géol. 44, 232 pp.

Newman, A.C.D. and Brown, G. 1966. Chemical changes during the alteration of micas. Clay Min., 6, 297-309.

Pedro, G. 1964. Contribution à l'étude expérimentale de l'altération géochimique des roches cristallines. Thèse,Faculté des Sciences de Paris. Ann. Agron., 344 pp.

482   A. Meunier

Perami, R. 1971. Contribution a l'étude expérimentale
    de la microfissuration des roches sous actions
    mécaniques et thermiques. Thèse, Faculté de
    Sciences de Toulouse.
Proust, D. and Velde, B. 1978. Beidellite crystallisa-
    tion from plagioclase and amphibole precursors :
    local and long-range equilibrium during weathering.
    Clay Min., 13, 199-209.
Richter, D. and Simmons, G. 1977. Microcracks in
    crustal igneous rocks : microscopy. In: The
    Earth's Crust. Amer. Geophys. Union, 149-180.
Rickwood, P.C. 1977. A technique for extracting small
    crystals from thin sections. Amer. Mineral.,
    62, 382-384.
Robert, M. 1971. Etude expérimentale de l'évolution
    des micas (biotites). Anns. Agron., 22, 155-181.
Seddoh, F.K. 1973. Altération des roches cristallines
    du Morvan : granites, granophyres, rhyolites.
    Etude minéralogique, géochimique et micro-
    morphologique. Thèse,Faculté des Sciences de
    Dijon. Mém. Géol. de l'Univ. de Dijon, 377 pp.
Simmons, G. and Richter, D. 1976. Microcracks in
    Rocks. In: R.G.J. Strens (Ed), The Physics and
    Chemistry of Minerals and Rocks, 105-137.
Stoops, G., Altemüller, H.J., Bisdom, E.B.A., Delvigne, J.,
    Dobrovolsky, V.V., FitzPatrick, E.A., Paneque, G.
    and Sleeman,J.R. 1979. Guidelines for the description
    of mineral alterations in soil micromorphology.
    Pédologie, 29, 121-135.
Tardy, Y. 1969. Géochimie des altérations. Etude des
    arènes et des eaux de quelques massifs cristallins
    d'Europe et d'Afrique. Thèse,Faculté des Sciences
    de Strasbourg, Mém. du Service de la Carte Geol.
    Alsace-Lorraine, 31, 199 pp.
Thenoz, B. 1966. Contribution à l'étude de la per-
    méabilité des roches et de leur altérabilité.
    Application à des roches granitiques. Thèse,
    Faculté des Sciences de Toulouse.
Trichet, J. 1970. Contribution à l'étude de l'altération
    expérimentale des verres volcaniques. Thése, Ecole
    Normale Paris, 152 pp.

Wallace, S.R. 1955.   Removal of mineral grains from
      their sections.   Amer.Mineral., 40, 927-931.
Wicks, F.J. and Zussman, J. 1975.   Microbeam X-ray
      diffraction patterns of the serpentine minerals.
      Can. Mineral., 13, 244-258.
Wilson, M.J. 1966.   The weathering of biotite in some
      Aberdeenshire soils.   Mineral. Mag., 35, 1080-1093.
Wilson, M.J. and Farmer, V.C. 1970.   Study of weathering
      in a soil derived from a biotite-hornblende rock.
      II. The weathering of hornblende.   Clay Min., 8,
      435-444.
Wilson, M.J. and Clark, D.R. 1978. X-ray identification
      of clay materials in thin sections.   J. Sed. Pet.,
      48, 656-660.
Wolff, R.G. 1967.   Weathering of Woodstock granite near
      Baltimore, Maryland.   Amer. J. Sci., 265, 106-117.
Wollast, R. 1967.   Kinetics of the alteration of
      K-feldspar in buffered solutions at low temperatures.
      Geochimica and Cosmochimica Acta, 31, 635-648.

THE MICROMORPHOLOGY OF PEAT

G.B. Lee

Department of Soil Science, University of Wisconsin,
Madison, Wisconsin

ABSTRACT
    Modern micromorphological studies of peat began
shortly after World War II. They were made in response
to a need for a better understanding of composition,
genesis and behavior of peatland soils, and the renewed
interest in pedological research at that time. In a real
sense they were an extension of macromorphological studies,
which began several decades earlier.
    The conceptual framework for micromorphological
research was in the substantial literature of peat and
other organic soil materials.
    Micromorphological techniques have been applied to the
study of soil formation in peat with the result that
we now possess a better understanding of the processes
involved and the kinds of soil produced from peat materials
in various environments.
    Several systems of classification and nomenclature
for organic soil materials have been designed in recent
years. Hopefully they will overcome some of the problems
of nomenclature and organisation present in earlier
systems.
    A new method of investigation, involving the use of a
Scanning Electron Microscope (SEM), holds promise for
greater application of peat micromorphology to basic soil
studies and to practical problems in agriculture, forestry
and engineering. There are also improved techniques in
optical microscopy, but the need for improved thin sections
persists.
    While much has been accomplished in the field of
peat micromorphology during the past 30 years, this method
of study has been applied to relatively few soils. It
is not used directly in the U.S. system of soil classification; few micromorphological studies of peat soils can
be found in computer searches of the literature.

INTRODUCTION

Micromorphological studies of peat and peatland soils are mainly confined to the second half of the twentieth century. With the exception of Kubiena's work (1938), few studies are reported earlier.

Most studies in the early post-war years appear to have been initiated as part of the resurgence of interest in pedological research following World War II. Micromorphological studies of peat followed naturally the macromorphological studies of peatlands which began several decades earlier. There was a need to know more about their composition, structure and fabric in order to place them properly in the new systems of classification being developed at that time. There was also a need to understand better their genesis, functioning and behaviour.

Previous studies had shown that peat could be studied microscopically with valuable results. Microscopy had been applied to the study of lignite and coal and was required for the identification of fossil plants used in the American classification (Thiessen, 1920) of these materials. Its use in the study of peat, therefore, was a natural step. In addition, both pre-war and early post-war studies of humus forms, by Kubiena (1938, 1953), had shown that it was possible to apply thin section technology to organic soil materials, and that the information obtained could usefully be applied to the classification of peatland soils.

NOMENCLATURE

Throughout the world, a variety of terms are used to describe peat soils and the land areas they cover. In the United States, 'peat', when used in a specific sense, refers to an organic soil or soil material containing identifiable plant remains; whilst 'muck' refers to an organic soil with no recognisable plant fragments (Soil Survey Staff, 1951). These terms are used commonly both in the scientific community and by lay people. However, the term peat is also used in a general sense to describe all soils in which the content of organic matter is so high as to dominate soil properties and behaviour. It is used in that sense here. U.S. soil scientists have also introduced the term Histosol (Histos: Gr. tissue, and solum: L. soil) to include all

organic soils (Soil Survey Staff, 1960, 1975). Histosols
formed under saturated conditions are subdivided according
to stage of decomposition and fibre content into three
classes, called Fibrists (high fibre content), Hemists
(moderate fibre content), and Saprists (low fibre content).
These terms are also used here.

Peat soils occur in peatlands, more commonly referred
to as swamps, marshes, bogs, moors, fens, muskeg, etc.
Natural peatlands support hydrophytic vegetation and are
generally saturated with water for long periods during
most years. The resultant anaerobic conditions favour
slow decomposition, resulting in an accumulation of vegeta-
tive debris which constitutes the parent material of peatland soils.

Peat soils have a worldwide distribution and consequently are formed in many kinds, or combinations, of
organic materials. Varying amounts of inorganic materials
are included. Characteristics of peat soils are determined
largely by the characteristics of their parent materials
and the kind and interaction of environmental factors
during soil formation.

PEAT MORPHOLOGY
History
Historically peatland soils have been described in
terms of their botanical composition, for example, sedge,
sphagnum, or woody peat. These terms had a general
usefulness and are still used in very general descriptions.

A.P. Dachnowski, a scientist employed by the U.S.D.A.
to study and classify American peat resources, was a
pioneer advocate of morphological studies. In an early
publication, Dachnowski (1919) gave credit to previous peat
researchers, mainly Europeans, noting however that they
"follow mainly botanical viewpoints and definitions".
He also observed that "few phases of botanical inquiry
have received as much attention as the development and
formation of peat deposits; yet information concerning
them appears to be little known and still less considered
in practice".

Dachnowski's scheme of classification (1919)
recognised four major genetic groups of peat materials,
namely Aquatic, Marsh, Bog, and Swamp. Both groups and
subgroups were described as to composition and appearance,

partly in morphological terms.

In a later paper (Dachnowski-Stokes, 1940) the author discussed peat and muck-forming processes in many of the same terms we use today. He also suggested the recognition of four basic macrostructural elements in peat soils, namely horizontal or laminated (i.e. platy), vertical (prismatic), blocky or fragmental, and granular. Five size classes and five grades of development were recognised for each type.

Many other scientists of that era from a variety of disciplines contributed to our present day theories on the genesis of peat soils and our overall knowledge of their morphology and composition. Among them was L. Von Post, who devised a scale for ranking peat materials in the field according to their relative degree of decomposition (Von Post and Granlund, 1926). In applying this scale, fresh samples are squeezed by hand and, on the basis of their behaviour and appearance, assigned a value ranging from H_1 (undecomposed, fibrous) to H_{10} (homogenous and decomposed). This test is very similar to the unrubbed-rubbed fibre content test used to classify Histosols in the United States today.

Waksman (1938) noted that many students of peat "overlook the fact that peat is a natural body, produced as a result of specific processes of transformation", and that "certain evidence points to the fact that some of the transformations whereby the plant materials are gradually changed into peat are microbiological in nature". Both concepts are basic to modern theories of soil forma-tion in peat.

Despite the research and publications of Dachnowski, von Post and others of their era, morphological studies of peatland soils were not common in the 1920s and 1930s. Little effort was being made to classify peat except very broadly on the basis of plant composition (e.g. moss vs herbaceous vs woody) and content of mineral matter, that is, as peat (low mineral content) or muck (high mineral content).

Modern morphologists have greater opportunities to carry out detailed studies; in fact, the many pressures on peatlands today require them to be studied in greater detail.

Morphological Studies
        Radforth (1956) studied the range of natural
structural conditions found in organic terrain in
Canada. Sixteen structural categories were recognised
and illustrated. Radforth and coworkers were particularly
concerned with the recognition of structural types that
could be correlated with stability, bearing capacity,
permeability, insulation value and other engineering
properties related to trafficability and road construction.
        Robertson (1962) discussed the origin and properties
of peat and its use in horticulture. While his primary
interest was directed toward proper classification and
grading for marketing, he did define several British
peat types, in part on their morphology. Kuiper and
Slager (1963) described the occurrence of distinct
prismatic and platy structures in Dutch Histosols,
correlating structure formation with a deep groundwater-
table.
        Boelter (1964, 1965) studied the water storage
characteristics and hydraulic conductivity of several
Minnesota peats as related to morphology and composition.
His results indicated that these properties varied
considerably according to pore-size distribution and bulk
density of histic materials, as influenced by fibre
content and stage of decomposition.
        Dolman and Buol (1968), studying the genesis of
Histosols on the lower coastal plain of North Carolina,
noted pedogenic structures. Frazier and Lee (1971)
studied the morphology and composition of three Wisconsin
Histosols classified as a Fibrist, Hemist, and Saprist,
respectively. It was noted that well-defined pedogenic
structures appeared to be restricted to the Saprist,
though not all Saprists possessed such structure. This
posed a problem in classification and suggested that
the Saprist category should be subdivided or redefined.

MODERN MICROMORPHOLOGICAL STUDIES
        Most micromorphological research on peat has been
done in western Europe. Lesser contributions have come
from the United States and other countries. Related work,
e.g. Barratt's studies of humus (1964, 1969), has contributed greatly. The author is unaware of any substantial
body of work on the micromorphology of peat by

scientists in the USSR although there is a considerable
volume of Russian literature devoted to palynology and
the botanical nature of peat.

Kubiena's Contributions
        W.L. Kubiena's work (1938, 1953) spanned the period
between pioneer and modern stages of pedomorphological
study. He identified humus forms by using converging
lines of evidence, taking into account chemical and
physical composition, location, specific biology,
phenological changes, and macro- and micromorphology.
His techniques and descriptions illustrating the micromorphological features of peat materials and other humus
forms provided later investigators with a technical
basis for further studies.

Application to Morphology and Genesis
        The centre of post-war activity for micromorpho-
logical research relating to the genesis of peatland
soils and their classification was in the Netherlands.
Soil scientists in that country (Pons, 1960; Van Heuveln
et al., 1960; Jongerius and Schelling, 1960) described
peat formation as a geogenetic process in which the parent
materials of organic soils were being accumulated. They
contrasted this process with the pedologic processes of
ripening (soil formation) which were initiated by peat
drainage and aeration. Ripening involved the physical
disintegration of plant parts and their biochemical
decomposition and biological granulation (moulding).
The latter, according to these investigators, caused the
formation of a distinct surface horizon as peat and other
material was repeatedly ingested and excreted by soil
fauna. Abundant nutrients, a near neutral pH, adequate
moisture, and aerobic conditions were noted as environmental factors that tended to encourage faunal activity
and accelerate the moulding process. Two kinds of
moulded horizons were recognised on the basis of micromorphological studies.
        One of these, the moder horizon, was described as
consisting mostly of faecal excrement from soil fauna
such as mites (Collembola), Diptera larvae and pot worms
(Enchytraeidae). Moder formation occurred in oligotrophic
peats containing very little clay, having a pH of 5 or

higher, and a C/N ratio greater than 17. This process
did not, however, involve the intimate binding of
organic and inorganic particles necessary to form in-
separable humus-mineral complexes as is the case in mull
formation (Kubiena, 1953). Jongerius (1957) and
Jongerius and Pons (1962) recognised two kinds of moder,
a small variety 25-60 µm diam. (after Collembola, Diptera),
and large moder 150-600 µm diam. (after Enchytraeidae).
Large and small moder together with fragments of plant
tissue and organic colloids sometimes formed large,
loosely aggregated granules called mull-like moder by
the Dutch workers.

Mull, in contrast to moder, was ascribed to intense
mixing and binding of organic materials with mineral
particles by organisms such as earthworms, Enchytraeidae,
and Julidae (Jongerius and Pons, 1962). The mull horizon
was described as consisting of faecal pellets approximately
2 mm diam. Mull formation was found to occur most commonly
under aerobic conditions in eutrophic or mesotrophic peats
which contained some clay and were near neutral in reaction.
Carbon/nitrogen ratios were less than 17. The size and
shape of mull aggregates could be altered by a change in
environment, for example, continued aerobic conditions
caused mull aggregates to coalesce into composites, while
prolonged anaerobic conditions apparently caused mull
aggregates to disperse into small granules.

Two important papers based on work in Germany were
published by Puffe and Grosse-Brauckmann (1963) and
Frercks and Puffe (1964). These reported investigations
into the decomposition of peat and the effects of drainage.

Building on this European experience, Langton and
Lee (1964) used micromorphological techniques and chemical
analyses to determine the composition and genesis of a
buried sapric horizon in a Wisconsin Histosol. The horizon
was characterised by its dark colour (N 2/) and strong
granular structure. Parts resembled moder or mull-like
moder. The latter consisted of moder and primary
constituents: disintegrated herbaceous material, black
fragments, and brown amorphous material. These micromorphological features suggested that the buried horizon
was a moder epipedon, a relict horizon formed at the
surface of the marsh in an earlier period of intense
faunal activity due to aerobic conditions brought about

by a lowered water-table.

Lee and Manoch (1974) investigated the micromorphological characteristics of major horizons in a
Wisconsin Saprist that had been farmed for 51 years,
following tile drainage. Several major horizons were
identified, each with definitive macro- and micromorphological characteristics:

The sapric surface horizon was dark coloured, granular,
and with little fibre. Thin section studies revealed a
high content of faunal aggregates. Sapric subsoil horizons
were also dark coloured. Primary peds were prisms and
blocks. The soil mass consisted mainly of black and brown
histic fragments in a brown amorphous matrix. Some
material was organised in cell-like micro-peds. Faunal
micro-aggregates were present in lesser amounts than in
the surface horizon. Fresh earthworm casts were noted
between structural elements and within the soil mass.
Apparent ped coatings suggested illuvial concentrations of
finely divided organic matter and/or mineral matter.
Pedogenic structure became coarse and weak with depth and
could not be discerned below the level of tile drains.
At that depth (c.100 cm) matted fibrous peat, saturated
with water, was found overlying limnic sediments.

The colour and macrostructure of the surface and
subsoil horizons, low fibre and high solubility in sodium
pyrophosphate suggested a high degree of decomposition
caused by pedogenic processes following drainage. The
micro-structure and other micro-features of the soil as
observed in thin section, provided the visual evidence
needed to confirm this hypothesis and explain the nature
of the processes involved.

Babel (1965, 1971) investigated the micromorphogical
characteristics of decaying organic matter in soils using
special microscopic techniques to identify plant materials
and their stage of decomposition. He also studied the
relationship of soil-forming processes to the morphology
of organic soil materials.

Bunting (1975) applied micromorphological techniques
to organic and mineral soils on Devon Island, Canada.
Results showed that humification of moss fibres and
plant debris varied from a little in buried moss peats to
much in a protosol of animal and bird perches, and to a
slightly lesser degree in earth hummocks.

Dinc et al. (1975) studied the morphological and chemical attributes of three peat soils: a sphagnum peat, a sphagnum-Eriophorum peat, and a Carex peat. The first two were characterised by a large amount of histons (Bal, 1973), while few were found in the third. Fungons were characteristic of the first soil, found in small amounts in the second, and absent from the third. Faunal activity and decomposition, as evidenced by faecal pellets, were slight in the first soil (mainly by Oribatid mites). The second soil was moderately decomposed, mainly by Oribatids but also earthworms and Enchytraeids. The third soil was highly decomposed by Diptera larvae and earthworms, and to a lesser extent by Oribatids. Rubbed fibre content and chemical data correlated well with morphological changes and the nature of the soil with depth.

Nomenclature and Classification of Organic Soil Materials
       The nomenclature and classification of organic soil materials has been the subject of numerous studies including Kubiena (1953), Jongerius and Schelling (1960), Jongerius and Pons (1962), Babel (1965), Barratt (1964, 1969), Bal (1973). Brewer's (1964) definitive treatise on fabric and mineral analysis of soils discussed the subject briefly, as related to faecal pellets. These and other articles were reviewed by Bullock (1974) who concluded that no systematic method for describing the micromorphology of humus forms or organic materials existed before 1965; that prior to that time there was no systematic descriptive approach on which to base lower levels of classification. Since that time several schemes had emerged.
       Brewer (1974) indicated that while genetic classifica-tion should not be entirely neglected, undue emphasis on genesis appeared in most systems, for example, the require-ment that mull and moder be formed by faunal activity. He suggested that "a prime need at present is a careful description of fabric features with a morphological classification for the accurate dissemination of observa-tional data between workers".
       Three relatively recent descriptive systems include those of Babel (1965), Barratt (1969), and Bal (1973). Babel's system was concerned mainly with the several stages of plant decay. It was based on morphological criteria such as staining, birefringence, and plant structure.

Barratt (1969) submitted a revised classification
and nomenclature of microscopic organic soil components
in which she attempted to avoid the confusion caused by
the existing nomenclature. Her scheme ordered soil
materials into five major classes according to their
organic and mineral components as follows:
    Skeletal Materials: 1. Humiskel - mostly undecom-
    posed organic residues; 2. Lithiskel - mineral
    grains and fragments.
    Plasmic Materials: 3. Humicol - decomposed organic
    colloids; 4. Mullicol - a mixture of organic and
    clay colloids; 5.Argillicol - clay colloids.
Barratt's main classes were subdivided according
to kind of constituents and microstructure, the latter
in terms of shape, size and arrangement (fabric) of
solid particles and voids.
    Bal's scheme of classification (1973) was designed
to provide a complete, basic descriptive system of organic
soil materials that could be used in conjunction with
Brewer's system (1964) for mineral soils. Bal described
all organic materials in a soil as part of a natural three-
dimensional entity which he named a Humon. He describes
a humon more specifically as "the collection of macroscopically and/or microscopically observable organic
bodies in soil which are characterised by a specific
morphology and spatial arrangement. The Humon profile
is the vertical section of a humon in which consecutive
exposed horizons represent the result and mode of
decomposition of organic material".
    Bal's system is detailed and comprehensive and
provides a systematic framework for the description and
organisation of organic soil materials. It is a
morphogenetic system and introduces a number of new terms.
This suggests that it may only be tested and used by
relatively few soil scientists, primarily those who are
serious students of peat or humus micromorphology.

METHODS OF INVESTIGATION
    The microscope is only one tool in an array available
to modern peat morphologists. Micromorphological studies
complement other field and laboratory studies in most
investigations, and several methods are usually required

to obtain the information needed. Cady (1974) noted
that the potential for interpreting and understanding
features seen in thin section is increased manyfold
according to the amount of other information on the
sample, on the profile and on the whole geomorphic and
biological setting.

Brewer (1964) describes techniques used in
micromorphological investigations and, in addition,
describes a number of other field and laboratory techniques
useful in obtaining the ancillary data useful in interpretation. Babel (1971) discussed technique of thin
section preparation and a variety of microscopic techniques,
including the use of a polarising microscope to detect
cellulose and the use of fluorescence. Kowalinski and
Kollender-Szych (1972) compared micromorphological,
field and physico-chemical methods for the determination of
decomposition rates of two organic soils. The results
obtained by these methods generally agreed. Micromorphological studies provided the best qualitative picture
of decomposition processes, but normal thin section methods
were unsatisfactory for examining highly decomposed
material. Jongerius (1974) described the use of a Zeiss
particle size analyzer and the Quantimet 720, an electro-
optical scanning device, for quantifying the area and
shape of objects under study. Levesque et al. (1978)
subjected samples of two peat soils to microscopic examination and pollen analysis. They concluded that the latter
procedure aided in the interpretation of micromorphological
features observed in thin section.

Preparation of Thin Sections
A persistent problem in micromorphological studies of
peat has been the preparation of satisfactory thin
sections. A wide variety of peat materials exist; they
must be stabilised, cut and polished before they can be
studied. Ideally, the impregnating material should be
colourless, isotropic, and hard at room temperatures.
The compound, and the technique used in preparation of the
thin section, should not cause volume change or disruption
of soil fabric. In the case of most organic soil materials,
shrinkage and disruption of natural fabric is a much
greater problem than it is with mineral soil samples.

Kubiena (1938) used Kollolith, which along with Canada balsam dissolved in Xylol,have been popular natural resins used in thin section preparation.   In recent years, a wide variety of polyester and epoxy resins have been used for both mineral and organic soil materials (Brewer, 1964; Grossman, priv. comm. ; Cent and Brewer, 1971).   Some are used with air-dried rather than oven-dried samples, thereby reducing the disruption of fabric and structure caused by heat and rapid drying.   Important developments have been made in the removal of water prior to impregnation of peats; in particular, the method of FitzPatrick and Gudmundsson (1978) using vapour from aqueous acetone has much to recommend it.

A few organic materials, e.g. unhumified moss peats, may be impregnated with paraffin or waxes and sliced on a microtome.   Mineral grains in all but the purest samples, however, restrict the use of this technique.   An aqueous agar solution may be used for some plant materials; gelatin was used by Minderman (1956).

Carbowax 6000, a high molecular weight polyethylene glycol compound, was used by Mitchell (1956) to impregnate moist mineral soil samples. MacKenzie and Dawson (1961) showed that moist samples of peat impregnated with Carbowax developed less shrinkage than oven-dried or lyophilised samples impregnated with a thermal-setting polyester resin.   When the latter was used to impregnate a sample of sedimentary peat, the organic matter shrunk into small particles and strands, in a matrix of diatoms. Good results were obtained when undried but similar samples were impregnated with Carbowax.

Langton and Lee (1965) experimented further with Carbowax, noting that it simplified the impregnating procedure because it replaced the water in wet samples of peat with little or no tissue shrinkage.   The main difficulty arose when it accumulated in the voids and pores of the sample, causing a blurry appearance.   Manoch (1970) improved the procedure further by mounting thin Carbowax impregnated slabs of peat on slides with Castoglas.   During this process the isotropic Castoglas melts and displaces much of the anisotropic wax in pores improving slide quality.

Despite the obvious advantages of Carbowax in preventing shrinkage, it is not a hard material, and

requires time-consuming hand preparation. All other
impregnating agents appear to have some disadvantages
also when used for peat samples. New impregnating
compounds and improved techniques of thin section
preparation are needed to encourage greater use of
microscopy in the study and classification of organic
soil materials.

Application of Scanning Electron Microscopy to Peat
     From time to time new tools are developed for
particular types of study, or tools designed for one
purpose are applied in a new way. Electron microscopy,
especially scanning electron microscopy, is a good
example.
     Dhowian (1978) used SEM to aid the characterisation
of peat materials and the effects of loading on peat
morphology. A magnification power of about 300 provided
clear viewing of individual fibres. Specimens used in
the investigation were air-dried and sampled in such a
manner that two surface orientations could be studied,
one parallel to the surface of the soil, the other
perpendicular to it. Four different peat materials were
analysed.
     Micrographs of undisturbed samples exhibited qualita-
tive differences in the size and kinds of fibres, their
orientation, degree of uniformity, and the relative
amounts and sizes of mineral particles present. It was
concluded that the four samples represented different
kinds of peat materials, as had been indicated by other,
conventional,methods of soil analysis.
     SEM micrographs made after loading showed that
consolidation had noticeable effect on the size,
spatial arrangement and orientation of peat fibres.
Fibres were reduced in size because of the expulsion of
water. Peat structure was noticeably finer with fibres
more closely packed and with thin, more or less parallel
edges. Their orientation was perpendicular to the applied
pressure. It was concluded that loading caused the
packing of peat fibres into a dense, stable arrangement.
     New tools, new techniques and more effective
application of tools and techniques presently used in
micromorphological studies have the potential for
greatly increasing our understanding of peat materials

and peatland soils.

REFERENCES

Babel, U. 1965. Die Ansprache von Pflanzenresten im
    mikroskopischen Präparet von Humusbildungen
    Zeitschrift Pflangenernährung, Düngun, Bodenkunde,
    109, 17-26.
Babel, U. 1971. Methods of investigating the micro-
    morphology of humus. In: H. Ellenberg (Ed),
    Ecological Studies. Analysis and Synthesis, Vol.2,
    Springer-Verlag, Berlin, 164-168.
Bal, L. 1973. Micromorphological Analysis of Soils.
    Soil Survey Papers 6. Neth. Soil Surv. Inst.,
    Wageningen, 174 pp.
Barratt, B.C. 1964. A classification of humus forms and
    micro-fabrics of temperate grassland. J. Soil Sci.,
    15, 342-356.
Barratt, B.C. 1969. A revised classification and nomen-
    clature of microscopic soil materials with particular
    reference to organic components. Geoderma, 2, 257-
    271.
Boelter, D.H. 1964. Water storage characteristics of
    several peats in situ. Soil Sci. Soc. Amer. Proc.,
    28, 433-435.
Boelter, D.H. 1965. Hydraulic conductivity of peats.
    Soil Sci., 100, 227-231.
Brewer, R. 1964. Fabric and Mineral Analysis of Soils.
    Wiley and Sons, New York. 470 pp.
Brewer, R. 1974. Some considerations concerning micro-
    morphological terminology. In: G.K. Rutherford (Ed),
    Soil Microscopy.   The Limestone Press, Kingston,
    Ontario, 28-48.
Bullock, P. 1974. The micromorphology of soil organic
    matter - a synthesis of recent research. In:
    G.K. Rutherford (Ed), Soil Microscopy. The Limestone
    Press, Kingston, Ontario, 49-65.
Bunting, B.T. 1975. Micromorphological observations of
    the interactions of biological activity and organic
    matter in Canadian High-Arctic soils. In: Proc.,
    Scientific Committee on Problems of the Environment.
    National Research Council, Ottawa.
Cady, J.G. 1974. Applications of micromorphology in soil
    genesis research. In: G.K. Rutherford (Ed), Soil

Microscopy. The Limestone Press, Kingston, Ontario, 20-27.

Cent, J. and Brewer, R. 1971. Preparation of thin sections of soil materials using synthetic resins. CSIRO, Div. of Soils, Tech. Paper 7, 18 pp.

Dachnowski, A.P. 1919. Quality and value of important types of peat material. U.S.D.A. Bull., 802, 40 pp.

Dachnowski-Stokes, A.P. 1940. Structural characteristics of peat and muck. J. Amer. Soc. Agron., 22, 389-399.

Dhowian, A.W. 1978. Consolidation effects on properties of highly compressible soils-peats. Ph.D. Thesis, Univ. of Wisconsin, Madison.

Dinc, U., Miedema, R., Bal, L. and Pons, L.J. 1976. Morphological and physico-chemical aspects of three soils developed in peat in the Netherlands and their classification. Neth. J. Agric. Sci., 24, 247-265.

Dolman, J.D. and Buol, S.W. 1968. Organic soils on the lower coastal plain of North Carolina. Soil Sci. Soc. Amer. Proc., 32, 414-418.

FitzPatrick, E.A. and Gudmundsson, T. 1978. The impregnation of wet peat for the production of thin sections. J. Soil Sci., 29, 585-587.

Frazier, B.E. and Lee, G.B. 1971. Characteristics and classification of three Wisconsin Histosols. Soil Sci. Soc. Amer. Proc., 35, 776-780.

Frercks, W. and Puffe,D.1964. Chemische und mikromorphologische Untersuchungen über den Zersetzungszustand auf entwässertem und verschieden hoch aufgekalktem Hochmoor unter Grünland sowie in unkultiviertem, vorentwässertem Hochmoor. Z. Kulturtech. Flurberein. 5, 149-171.

Jongerius, A. 1957. Morfologishe on derzoekingen over de boemstructure. Bodemkundigl studies No.2. Wageningen, The Netherlands.

Jongerius, A. 1974. Recent developments in soil micromorphology. In: G.K.Rutherford(Ed),Soil Microscopy. The Limestone Press, Kingston, Ontario, 67-83.

Jongerius, A. and Schelling, J. 1960. Micromorphology of organic matter formed under the influence of soil organisms, especially soil fauna. Trans. 7th Int. Congr. Soil Sci., Madison, II, 702-710.

Jongerius, A. and Pons, L.J. 1962. Soil genesis in organic
    soils. Boor en Spade, XII. Soil Surv. Inst.,
    Wageningen, The Netherlands, 156-168.
Kowalinski, St. and Kollender-Szych, A. 1972. Micro-
    morphological and physico-chemical investigations
    on the decomposition rate of organic matter in some
    muck soils. In: St. Kowalinski (Ed), Soil Micro-
    morphology. Warsaw, Poland, 144-155.
Kubiena, W.L. 1938. Micropedology. Collegiate Press,
    Ames, Iowa.
Kubiena, W.L. 1953. The Soils of Europe. Thomas Murby
    and Co., London.
Kuiper, F. and Slager, S. 1963. The occurrence of distinct
    prismatic and platy structures in organic soil
    profiles. Neth. J. Agric. Sci., 11, 418-421.
Langton, J.E. and Lee, G.B. 1964. Characteristics and
    genesis of some organic soil horizons as determined
    by morphological studies and chemical analysis.
    Proc., Wis. Acad. Sci., Arts, and Letters, 53,
    149-157.
Langton, J.E. and Lee, G.B. 1965. Preparation of thin
    sections from moist organic soil materials. Soil
    Sci. Soc. Amer. Proc., 29, 221-223.
Lee, G.B. and Manoch, B. 1974. Macromorphology and
    micromorphology of a Wisconsin Saprist. In: A.R.
    Aandahl (Ed), Histosols, their Characteristics,
    Classification, and Use. Soil Sci. Soc. Amer.
    Spec. Publ. 6, 47-62.
Levesque, M., Richard, P. and Dinel, H. 1978. Analyse
    pollinique et micromorphologique de deux tourbes
    du sud-ouest du Quebec. Can. J. Soil Sci., 58,
    525-528.
Mackenzie, A.F. and Dawson, J.E. 1961. The preparation
    and study of thin sections of wet organic soil
    materials. J. Soil Sci., 12, 142-144.
Manoch, B. 1970. Micromorphology of a Saprist. M.S.
    Thesis, University of Wisconsin, Madison.
Minderman, G. 1956. The preparation of microtome sections
    of unaltered soil for the study of soil organisms
    in situ. Plant and Soil, 8, 42-48.
Mitchell, J.K. 1956. The fabric of natural clays and
    its relation to engineering properties. Proc. High-
    way Research Board, 35, 693.

Pons, L.J. 1960. Soil genesis and classification of
     reclaimed peat soils in connection with initial
     soil formation. Trans. 7th Int. Congr. Soil Sci.,
     Madison. IV, 205- 211.
Puffe, D. and Grosse-Brauckmann, G., 1963.   Mikromor-
     phologische Untersuchungen an Torfen. Z. Kultur-
     technik und Flurberein.,4, 159-188.
Radforth, N.W. 1956. Range of structural variation in
     organic terrain. Tech. Memo 39. Nat. Res. Council,
     Ottawa.
Robertson, R.A. 1962. Peat: Its origin, properties, and
     use in horticulture. Scientific Horticulture, 16,
     42-52.
Soil Survey Staff. 1951. Soil Survey Manual. U.S.D.A.
     Handbook 18. U.S. Govt. Print. Office, Washington,
     D.C.
Soil Survey Staff. 1960. Soil Classification, 7th
     Approximation. U.S.D.A. Soil Conservation Service.
     Washington, D.C.
Soil Survey Staff. 1975. Soil Taxonomy. U.S.D.A.
     Handbook 436. U.S. Govt. Print. Office, Washington,
     D.C.
Thiessen, R. 1920. Compilation and composition of
     bituminous coal. J. Geol., 28, 185-209.
Van Heuveln, B., Jongerius, A. and Pons, L.J. 1960.
     Soil formation in organic soils. Trans., 7th Int.
     Congr. Soil Sci., Madison, 195-204.
Von Post, L. and Granlund, E. 1926. Sveriges Geologiska
     Undersokninz, 335, 29.
Waksman, S.A. 1938. Humus. Williams and Wilkins Co.,
     Baltimore.

THE CHRONOSEQUENCE OF PEDOGENIC PROCESSES IN FRAGLOSSU-DALFS OF THE BELGIAN LOESS BELT

R. Langohr and G. Pajares

University of Ghent, Ghent, Belgium

ABSTRACT

The chronological ordering of various features in a Fraglossudalf of Belgium has been attempted using macroscopic, mesoscopic and microscopic levels of observation. Four types of degradation feature are present: (B) tongues, A'2gx tongues, B'1gx tongues and skeletans and skeletspots. Evidence is provided to show that the B'1g tongues containing numerous disturbed illuvial clay bodies are the oldest, followed in more recent times by the development of the A'2g tongues, a fragipan, skeletans and skeletspots, and finally the (B) tongues. The formation of skeletans and skeletspots is associated with turbulent water flow probably in a periglacial environment. The tongues have a morphology consistent with root galleries.

INTRODUCTION

A detailed morphological study has been made of the pedogenetic processes concerned with the 'degradation' of the Bt horizon in Fraglossudalfs (Fig. 1). The profiles are situated on flat to slightly sloping uplands under forest. The annual rainfall is 750 mm and the soils are well drained. The parent material consists of several metres of Weichselian loess originally containing some 10-12% free $CaCO_3$. The depth of decalcification is at present about 2.5m. The soils were previously described by Louis (1959). A fragipan is considered to be present (Langohr and Van Vliet, 1981) and this investigation suggests that the whole profile can be described as a bisequum (Fig. 1), similar to that in soils with fragipans in the Belgian Ardennes (Langohr and Van Vliet, 1979). The lower part of the sequum, including the albic (A'2gx), argillic (B'2t) and fragipan (A'2gx - B'1gx) horizons, seems to relate to past pedogenetic processes (Langohr and Vermeire, 1982).

The object of this study has been to establish a
link between the field study (macroscopy), microscopic
observation of thin sections (microscopy) and a detailed
description of undisturbed soil fragments under the
stereomicroscope (mesoscopy). Using these observations
the chronological ordering of the various processes has
been obtained.

DEGRADATION PROPERTIES
        The field study revealed at least four degradation
features of the B'2t horizon. Three are tongues, in the
(B), the A'2gx and the B'1gx horizons (Fig. 1), the
fourth are skeletans situated along some of the ped faces
and tubular pores, and small (up to 5 mm diameter),
completely bleached, colloid-leached pockets of the
matrix, termed skeletspots. The following chronosequence can be proposed on the basis of the relative
positions of these features. From the oldest to the
most recent, development involves:
1.   B1g tongues, after the major period of clay migration,
2.   A2g tongues, before the development of the fragipan
     and the pseudogley,
3.   skeletans, after the genesis of the fragipan and
     probably synchronously with the pseudogley,
4.   (B) tongues, corresponding to the present active
     processes (mainly bioturbation).

OBSERVATIONS AND DISCUSSION
        The (B) tongues pass through the underlying A'2gx
horizons as far as the lower limit of the B'22t (Fig. 1).
The tongues have a morphology consistent with root
galleries. Nearly all present day biological activity
is in the tongues. They are not bleached and probably
for this reason have not been recognised before in these
soils. The mesoscopic observation shows that most of
the pores are tubular pores (fine root galleries)
(Fig. 2a).
        The compact fragipan, through which rare roots pass,
has a characteristic lenticular platy structure
(Fig.2 b, c, d). The pan includes the A'2g (Fig. 2b),

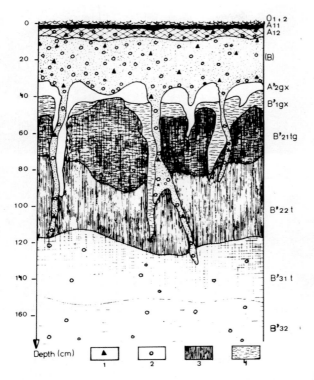

Fig. 1.   Schematic representation of the irregular
          horizon sequence of the Fraglossudalfs of the
          Belgian loess belt.

(Key : 1-charcoal fragments;   2-living roots;
        3-clay illuviation;   4-pseudogley mottling).

the B'g (Fig. 2 c,d), and often also the upper part of
the B'21tg . The thin sections (Fig. 2b,d) show (1)
relatively low intrapedal porosity, (2) the presence
of discontinuous fissures along the plates, (3) the
presence of skeletans on top of the plates, and (4)
that the fissures and associated skeletans cross pre-
existing illuvial clay bodies (centre of Fig 2d).
     Skeletans are very common on the upper face of
the lenticular plates of the fragipan (Fig. 2b, c,d)
and occur along many of the vughs (Fig. 3a) and fissures
(Fig. 3b, d) in the B'2t horizon.   Skeletspots, which

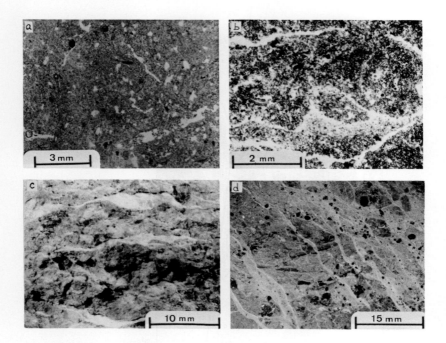

Fig.2. (a) (B) horizon with very high porosity
        (mainly tubular pores); (b) A'2gx horizon:
        upper half showing discontinuous fissure along a
        lenticular plate, lower half showing a well
        developed skeletan on the upper surface of a
        plate; (c) Mesoscopic view of the lenticular
        platy structure of a B'1gx horizon. Skeletans
        cover the upper face of the plates; (d) B'1gx
        horizon with skeletans nearly continuous on the
        oblique lenticular plates, centre of the
        figure shows disturbed clay cutans crossed by
        fissures and skeletans.

are not associated with the present day porosity, are
scattered through the B'1gx, B'21tg and B'22t horizons,
but dominantly in the first two horizons. Thin section
studies indicate that the skeletans and skeletspots,
(1) completely lack colloidal material (which differentiates them from the bleached A'2gx tongues which still
contain 7-10% clay), (2) mostly consist of well sorted

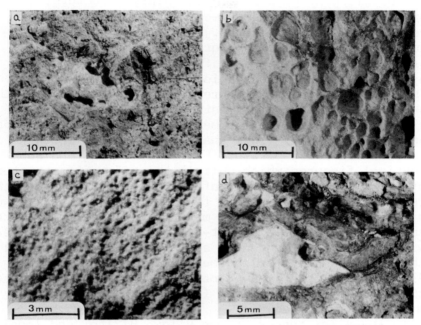

Fig.3. (a) B'21tg horizon showing skeletan associated
       with three tubular pores;   (b) Area of well
       developed micro-potholes on top of a thick
       skeletan situated along a vertical fissure of the
       B'22t horizon;   (c) Micropitting on a skeletan
       situated on top of a lenticular plate in the
       B'1gx horizon;   (d) A skelet-band and associated
       clay cutan (to the right of the skelet-band)
       along a vertical fissure of the B'22t horizon.

skeletal grains (size range 10-50 μm), sometimes with
graded bedding, and (3) locally include fragments of
broken cutans. The hypothesis is advanced that skeletans
are the consequence of turbulent water flow. This process
was probably active in periods of cyclic deep freezing and
thawing of the soil. At the onset of the thawing at the
surface, quantities of water could flow between the ice
needles situated in the fissures and cavities of the
still frozen subsoil. Such an hypothesis is supported by
field observation made by the senior author who observed
in one single freeze-thaw cycle in massive loess deposits

the development of a lenticular-platy structure, the
presence of numerous ice needles in the fissures and
cavities, and the presence, after thawing, of a thin
layer of leached skeletal grains on the plates.   The
process of turbulent water flow is further supported
by the mesoscopic observations of the Fraglossudalfs.
These show the localised presence at the surface of the
skeletans of (1) small (mostly less than 5 mm diam.)
closed depressions called here micro-potholes (Fig. 3b)
and (2) very small (less than 300 μm diam.) holes
called here micro-pitting (Fig. 3c).   Along the sub-
vertical fissures which delimit the prismatic structure
of the B'2t the skeletans often occur as irregular, thick
(a few millimeters up to a few centimeters) bands (skelet-
bands) (Fig. 3d), oriented more or less parallel to the
soil surface.   These skelet-bands are often associated
with clay and/or manganese and/or iron cutans.   Locally
the skelet-bands overlap the other cutanic features.
All these observations point to an erosion - sedimentation process which Kubiena (1938, p.163) termed
'microscopic-erosion'.
     The A'2gx tongues cross B'lgx, B'2ltg, and locally
even B'22t horizons.   These tongues have a macromorphology of root galleries.   They contain 7-10% clay
which is rather bleached.   Microscopic observations
show that broken clay cutans, unrelated to present
porosity, and small (< 1 cm diam.) fragments of the
Bltg horizon occur fairly commonly in the horizon.   More
recent undisturbed bleached argillans are present in the
lower part of a few tongues.
     B'lgx tongues in thin section contain numerous
disturbed argillans.   This disturbance occurred before
the development of the lenticular platy structure of
the fragipan as evidenced by the presence of fissures
and associated skeletans which cross previously broken
cutans.

CONCLUSIONS
     The mesoscopic and microscopic observations support
the chronosequence of pedogenetic processes proposed
from field observations:
1.   The Blg tongues are the oldest degradation feature.
They appear to have developed after the period of major

clay migration in the profile judging from the fact
that all illuvial clay bodies in the tongues are extensively disturbed.
2.   The presence of small broken fragments of Blg tongues
in the A'2g tongues, suggest a later development for the
latter.
3.   The fragipan developed after the genesis of both the
B'lg and A'2g tongues.
4.   Skeletans developed either during, or most probably
after, the genesis of fissures and cavities of both the
fragipan and the B'2t horizons. Their period of
formation seems to be more or less synchronous with the
development of the pseudogley. The genesis of the
skeletans seems to be an erosion-sedimentation process
associated with turbulent water flow, probably in a
periglacial environment. The clay cutans along the
fissures and large tubular pores of the lower part of
the B'2t and of the upper part of the B3 (B3lt), seem
to be linked to this process.
5.   Skeletspots possibly correspond to a similar period
of turbulent water flow probably occurring before
that described under 4.
6.   The (B) tongues are the most recent and reflect the
only active process of degradation of the Bt horizon
in these profiles. They correspond to the few areas
where tree roots succeed in passing through the fragipan.
Nearly all the present biological activity is situated
in these tongues and there is no evidence of recent clay
migration. The genesis of these tongues seems to be
purely mechanical.

REFERENCES
Kubiena, W.L. 1938. Micropedology. Collegiate Press,
       Ames, Iowa. 243 pp.
Louis, A. 1959. Verklarende tekst bij het Kaartblad
       Uccle 105 W. Soil Map of Belgium, IWONL-CVB,
       Ghent, 90 pp.
Langohr, R. and Van Vliet, B. 1979. Clay migration in
       well to moderately well drained acid brown soils
       of the Belgian Ardennes. Morphology and clay
       content determination. Pédologie, 29, 367-385.

Langohr, R. and Van Vliet, B. 1981. Properties and
    distribution of Vistulian permafrost traces in
    today surface soils of Belgium, with special
    reference to the data provided by the soil survey.
    Biul. Peryglac., 28, 137-148.
Langohr, R. and Vermeire, R. 1982. Well drained soils
    with degraded Bt horizon in loess deposits of
    Belgium; relationship with paleoperiglacial
    processes. Biul. Peryglac., 29, 203-212.

EVIDENCE OF DISTURBANCE BY FROST OF PORE FERRI-ARGILLANS IN SILTY SOILS OF BELGIUM AND NORTHERN FRANCE

B. Van Vliet[1] and R. Langohr[2]

[1] Centre de Pédologie Biologique du C.N.R.S.
Vandoeuvre, France
[2] State University of Ghent, Ghent, Belgium

ABSTRACT
     The systematic microscopic study of silty soils with Bt horizons in Belgium and northern France reveals that most of the pore ferri-argillans have undergone weak to severe perturbations. Cracking, cleaving, complete detachment from the s-matrix and traces of effects of stress are recognized. Frost action seems the most plausible cause of these disturbances. This hypothesis is supported by the presence in these soils of one or more of the following features : (1) a platy structure with discontinuous fissures with smooth walls along the plates; (2) a fragipan; (3) matrans, siltans or skeletans along tubular pores and fissure walls; (4) silty cappings on stone fragments; (5) vertically-oriented stone fragments; (6) clear traces of cryoturbation. It is concluded that most of the ferri-argillans developed before the last cold period(s) of the Weichsel glaciation.

INTRODUCTION
     A microscopic study of soils with Bt horizons in Belgium and northern France reveals that most pore clay cutans are broken, fissured, or at least distorted, by stress. These properties of the cutans may be attributable to extreme drying of the samples and to the preparation of the thin sections. However, the degree of disturbance appears to be related to position in the profile; it is most intense in the Bl, very important in the B2t, and strongly decreases in the B3t (Fig. la). Furthermore, numerous observations of soils with well developed Bt horizons in subtropical and Mediterranean areas show the absence of such disturbance. Yet these soils are subject to more intensive wetting-drying cycles than

those of western Europe and generally have a higher
proportion of swelling clays. The problem is to
determine the cause of disturbance of the clay-illuvial
bodies in these soils in Belgium and France.

RESULTS

　　Most of the soils with disrupted cutans have a weak
to strong platy structure. Under the microscope this
is seen as discontinuous smooth-walled fissures, more or
less parallel to the soil surface. This structure and
the associated porosity directed attention towards
desiccation or frost as possible mechanisms. Considering
the depth to which these fissures occur (sometimes up to
120 cm) and the fact that most roots are in the eluvial
horizons (upper 30-50 cm of the soil), the hypothesis of
desiccation by evapo-transpiration and the concomitant
development of shrinkage fissures is hardly tenable.

　　Fig. 1. (a) Undisturbed pore ferri-argillans in the
C1 horizon. Fraglossudalf of the loess belt.
PPL; (b) Traces of ice lenses (discontinuous
smooth fissures) in the B3x horizon of a Fragiudult of the High Ardennes. PPL; (c,d) Evidence of
cleavage and stress in the B2tx horizon of a
Fragiudalf of the loess belt. PPL (c) and XPL (d);
(e) Polysequum of voids and clay illuviation in
the B3x of a Fragiudult of the High Ardennes. PPL;
(f) Traces of freeze-thaw process in the B1x
horizon of a Fraglossudalf of the loess belt.
Fissures filled by sorted skeleton grains and
papules. PPL; (g) Traces of freeze-thaw process
in some horizon as in (f). Note the already disturbed illuvial clay bodies. PPL; (h) Evidence
of cryoturbation. Frost-uplifted shale fragments
in the B3x horizon of a Fragiudalf of the loess
belt. PPL.

A = agricutan;　C = clay illuviation;
P = papule;　S = sorted infilling;　V = void.

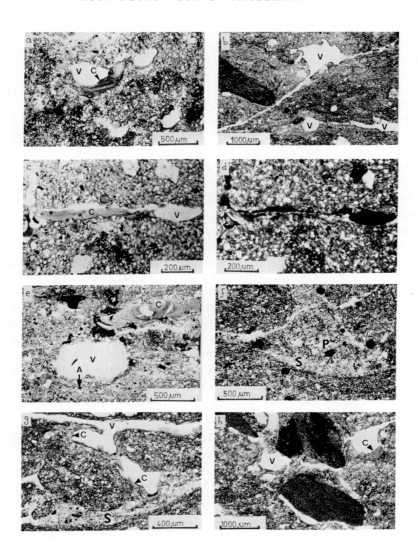

Fig. 1.

In the soils studied, these fissures are often
found in conjunction with a rather well-developed fragipan
(Langohr and Van Vliet, 1979; Van Vliet and Langohr,
1981). All these fissures are attributable to ice
veins (segregation ice: FitzPatrick, 1956; Van Vliet,
1976). In fragipans, illuvial clay bodies are
situated on only few of the fissure walls probably
because of discontinuity of the pores (Van Vliet, 1976,
1980a) (Fig.1b); along the prismatic structure faces
of the fragipans, the clay cutans are better developed,
and diffuse into the prisms following the fissures;
they are commonly disrupted (Caillier, 1977; Van Vliet
and Langohr, 1981). In profiles without a clear
fragipan, illuvial bodies are disturbed, often by clean
or skeletan-filled fissures.
The following features were observed in most profiles :
1. Ferri-argillans (Brewer, 1964),coating tubular pores
or fissures, may be cracked or even cleaved when a
second generation of fissures with smooth walls follows
the pre-existing one; locally this causes some lateral
shift (Langohr and Pajares, 1981).
2. Ferri-argillans may become detached from pores or
fissure walls and appear to be surrounded by a network
of fissures with smooth walls (Fig. 1c, d). These
dislocated argillans show the effect of stress (Fig. 1c,
d, e).

3. The second generation of fissures (see 1) resembles
porosity left by ice segregation. Associated with these
fissures, leached accumulations of mineral grains
are observed. These are often described as a degradation
property of the Bt horizons (Langohr and Pajares, 1981).
Under the microscope they appear as thin, sometimes
graded, accumulations of leached mineral grains,
often of a particular size, usually coarse silt (Fig.
1f). This feature is believed to result from flushing
of water during thawing in soils with segregated ice
lenses (Van Vliet, 1976, 1981) or fine ice needles
situated in tubular pores and fissures (Langohr and
Pajares, 1981) (Fig. 1g). Field observations show that
this process leaves observable traces after only one
single freeze-thaw cycle; in experiments it occurs
frequently in ice-rich silts (Dumanski, 1964;
J.P. Coutard, pers. comm.) A great number of these

cycles amplify   the development and grading of the
skeletans.   The occurrence of this process in a boreal
environment is well known (Dumanski, 1964; Fedorova
and Yarilova, 1972; FitzPatrick, 1956, 1974; Romans
et al., 1966).
4. Under forest,currently active drainage pores and
cracks lack recent illuvial bodies of well-oriented
fine clay and only matrans, siltans or skeletans
partially cover the pore walls (Fig. 1e).
5. Under forest most of Bt horizons are devoid of active
faunal turbation.
6. In soils with some stones (Ardennes region) other
features typical of freeze-thaw cycles are present
including (1) silt-clay cappings on stone fragments
(FitzPatrick, 1956), (2) vertically-oriented shale and
slate fragements, and (3) features due to intense cryoturbation. These properties are recognisable both
macroscopically and microscopically. They are most
intense in the lower part of the A'2 and in the upper
B't horizon (Langohr and Van Vliet, 1979) (Fig. 1h).
Even old bioturbation channels may be affected by
these frost features.

DISCUSSION
     The following stages in the development of silty
soils with some free $CaCO_3$ in the original parent
material are proposed :
1. Decarbonatation, followed more or less rapidly by
clay eluviation/accumulation from at least the
beginning of the Tardiglacial (large quantities of $CaCO_3$
existed in groundwater around 15,000 y B.P.) (Haesaerts,
1973; Van Vliet, 1980a).
2. Before the end of the Tardiglacial there were one
or more periods cold enough to freeze the whole Bt
horizon. A discontinuous permafrost is known to have
existed in High Belgium during the Youngest Dryas
(Pissart et al., 1979) and it is reasonable to suppose
that the rest of the country was influenced by intensive
seasonal frost in this period. Other periods with
particularly cold winters should also not be excluded,
namely the cold fluctuations of Piottino ($\pm$ 7000 y B.P.,
Woillard, 1974) and the Little Ice Age.
     Today these profiles are nearly always bisequal

(Langohr and Van Vliet, 1979). A colour and/or
structure (B) horizon or a weakly developed podzol form
the upper sequum; a Bt horizon, with or without
fragipan corresponds to the lower one. The lower sequum
is now inactive judging by (1) the absence of undisturbed
clay coatings and (2) the aluminisation and hence
flocculation of the clay (Van Vliet, 1980b). It can be
concluded that in nearly all the soils with Bt horizons
in Belgium and northern France, clay migration took place
mainly in the Interpleniglacial. This is in agreement
with the observations of Hoeksema and Edelman (1960) in
the Netherlands who found that the major clay migration
in the soils of the terraces of the Meuse river occurred
before the Holocene. In the loess deposits of Belgium
and northern France the clay-illuvial bodies have been
disturbed by frost during the last cold fluctuation(s)
of the Weichsel. It is thus concluded that the impact
of frost associated with the complex paleoclimatic
events of the Tardiglacial must be taken into
consideration for the understanding of the genesis and
evolution of soils in Belgium and northern France.
Periglacial processes appear not only to have affected
the general profile morphology but also have left
numerous microscopic traces, even at the level of
pore ferri-argillans which are considered here to be
relict features.

REFERENCES
Brewer, R. 1964. Fabric and Mineral Analysis of Soils.
    John Wiley & Sons, London, 470 pp.
Caillier, M. 1977. Etudie chronoséquentielle des sols
    sur terrasses alluviales de la Moselle. Genèse et
    évolution des sols lessivés glossiques. Thèse de
    spécialite en Pédologie, Univ. de Nancy I, 87 pp.
Dumanski, J. 1964. A micropedological study of eluviated
    horizons. M.S. Thesis, Saskatoon University, 94 pp.
Fedorova, N.N. and Yarilova, A. 1972. Morphology and
    genesis of prolonged seasonally frozen soils in
    Western Siberia. Geoderma , 7, 31-43.
FitzPatrick, E.A. 1956. An indurated horizon formed by
    permafrost. J. Soil Sci., 7, 248-254.
FitzPatrick, E.A. 1974. Cryons and Isons. Proc.

North of England Soils Disc. Gp, 11, Penrith
1974, 31-43.

Haesaerts, P. 1973. Contribution à la stratigraphie des
dépôts Pleistocènes Supérieur du Bassin de la
Haine. PhD Thesis, Free University of Brussels,
395 pp.

Hoeksema, K.J. and Edelman, C.H. 1960. The role of
biological homogenization in the formation and
transformation of gray-brown podzolic soils.
Trans. 7th. Int. Congr. Soil Sci., Madison, U.S.A.,
IV, 402-405.

Langohr, R. and Van Vliet, B. 1979. Clay migration in
well to moderately well drained acid soils of the
Belgian Ardennes. Morphology and clay content
determination. Pédologie, 29, 367-385.

Langohr, R. and Pajares, G. 1983, The chronosequence
of pedogenic processes in Fraglossudalfs of the
Belgian loess belt. In : P. Bullock and C.P.
Murphy (Ed), Soil Micromorphology. A.B. Academic
Publishers, Oxford, 503-510.

Pissart, A., Bastin, B. and Juvigné, E. 1979. Les traces
de buttes périglaciaires (Pingo ?, Palses ?) des
Hautes Fagnes. Biul. Peryglac., 29, 333-340.

Romans, J.C., Stevens, J.H. and Robertson, L. 1966.
Alpine soils of northern Scotland. J. Soil Sci.,
17, 184-199.

Van Vliet, B. 1976. Traces de ségrégation de glace en
lentilles associées aux sols et phénomènes
périglaciaires fossiles. Biul. Peryglac., 26,
42-55.

Van Vliet, B. 1980a. Correlations entre fragipan et
permagel. Application aux sols lessivés glossiques.
Notes et comptes-rendus du groupe de travail 'Régio-
nalisation du Périglaciaire', 5, 9-22.

Van Vliet, B. 1980b. Approche des conditions physico-
chimiques favorisant l'auto-fluorescence des minéraux argileux. Pédologie, 30, 369-390.

Van Vliet, B. 1981. Structures et microstructures formées
par la glace de ségrégation. In : H. French and
D. Judge (Ed), Proc. 4th Canadian Permafrost Conf.
Calgary.

Van Vliet, B. and Langohr, R. 1981. Correlation between
    fragipans and permafrost with special reference to
    Weichsel silty deposits in Belgium and northern
    France. Catena, 8, 137-154.
Woillard, G. 1974. Recherches palynologiques sur le
    Pléistocène dans l'Est de la Belgique et des
    Vosges Lorraines. Acta Geographica Lovaniensia,
    14, 168 pp.

AMOUNT, CHARACTERISTICS AND SIGNIFICANCE OF CLAY ILLUVIATION FEATURES IN LATE-WEICHSELIAN MEUSE TERRACES

R. Miedema, S. Slager, A.G. Jongmans and Th. Pape

Department of Soil Science and Geology,
Agricultural University,
P.O. Box 37,
Wageningen, The Netherlands

ABSTRACT
    A micromorphological study of 8 soils on the three Late-Weichselian terrace levels of the Meuse revealed that, contrary to the conclusion of van den Broek and Maarleveld (1963), these levels cannot be separated on the basis of subsoil colour or a quantification of clay illuviation phenomena. The absence of illuviated clay in soils of the Holocene terrace provides a means for separation of this terrace from the others.
    The various terrace levels can only be separated unambiguously on the basis of elevation. Particle-size distribution offers a possibility of differentiating Terrace III from I and II in accordance with van den Broek and Maarleveld (1963). Conclusions based on textural differences within the profiles are hazardous as clay illuviation need not coincide with these differences.
    The present hydrological position of the soils is responsible for modifications of the morphology of the clay illuviation features, which can be attributed to subsequent processes of soil formation as described here.

INTRODUCTION
    The Late-Weichselian Meuse terraces were described, dated and characterised by van den Broek and Maarleveld (1963). They presented a terrace map based upon geomorphology and characteristic longitudinal gradients of the fluvial Meuse terraces. They dated the various terraces using palynological and archaeological evidence and the occurrence of an Allerød soil in overlying coversand. They claimed that the soils on these terraces could be distinguished by means of characteristic particle-size distributions and the decreasing expression

of argillic horizons with decreasing age of the terraces.
The latter conclusion was mainly based on field evidence
of textural differences within profiles and colour
differences in the Bt horizons on the various terraces.
The presence of illuvial clay was checked in thin
sections.

        A soil survey training project near Broekhuizen
has provided a considerable amount of detailed information
concerning landscape and soils. The landscape development
and a comprehensive review of soil formation in the
surveyed area will be published in a forthcoming paper.
The present paper presents the results of the particle-
size analyses and micromorphology of selected soils with
special reference to the amount, characteristics and
significance of clay illuviation features in relation
to terrace level and drainage position.

RESULTS AND DISCUSSION

The Terrace Map (Fig. 1)
        The surveyed area was the subject of a detailed
geomorphological study. Measured heights at intervals of
10 cm on a scale 1 : 10,000 were corrected for the
thickness of coversand, river dunes, plaggen epipedons
and for human activities such as excavations and
sanding (sand introduced by human activity). This
allowed reconstruction of the topographic levels of
the terraces in the area, taking into account the
longitudinal gradients according to van den Broek and
Maarleveld (1963) (Fig. 1). The terraces are
designated I, II, III and IV with decreasing height
and age (Table 1). This corresponds to the
division of van den Broek and Maarleveld, except for an
area east of the Meuse which is considered to belong to
level II instead of I, based on its elevation.

        The separation of terraces I, II and III can be
easily made on elevation, but the separation between
III and IV is more difficult. van den Broek and
Maarleveld (1963) drew the same conclusion, and
elevations of the various levels are similar to theirs
(Table 1).

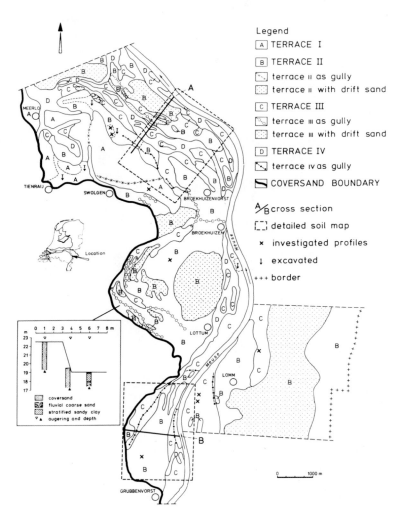

Legend

A	TERRACE I
B	TERRACE II
	terrace II as gully
	terrace II with drift sand
C	TERRACE III
	terrace III as gully
	terrace III with drift sand
D	TERRACE IV
	terrace IV as gully
	COVERSAND BOUNDARY

A/B cross section

detailed soil map

×   investigated profiles

↓   excavated

+++ border

Fig. 1.   Terrace map of surveyed area.

The Selected Soils

The investigated soils characterise the three oldest terraces and represent heavier textured variants. They generally range from well to imperfectly drained. The proportion of imperfectly drained soils is lowest on Terrace I (< 30%), increases to about 50% on Terrace

Table 1. Site and soil data.

Age	van den Broek and Maarleveld (1963) Elevation (Broekhuizen) m + NAP *	Terrace level	Present study Elevation (Broekhuizen) m + NAP	Soil name	Altitude m + NAP	Drainage condition	Subsoil colour Munsell colour moist	Profile illuviation index % cm	Thickness Bt horizon cm
Pre-Bølling	20.2	I	20 ± 1	Looiveld	19.1	well drained	7.5YR5/6	360	75
Allerød	17.8	II	18 ± 1	Gun 2	18.1	well drained	10YR4.5/6	460	90
				Gun 1	18.3	well drained	5YR4/4	450	100
				Grubbenvorst	17.8	imperfectly drained	7.5YR5/8 + 10YR7/2	250	60
				Roamweg	18.4	imperfectly drained	7.5YR5/8 + 2.5YR7/1	460	125
Late Dryas/	15.9	III	16 ± 1	Lomm 3	15.8	well drained	5YR4/4	640	125
Preboreal				Lomm 2	15.8	mod. well dr.	5YR4/4	340	95
				Lomm 1	16.0	poorly drained	7.5YR7/1 + 5YR6/4	310	70
Holocene	15.2	IV	15 ± 1	————					

* Dutch Ordnance Datum

II, and is over 70% on Terrace III.

Particle-size Distributions and Subsoil Colour
    On all three terraces the sand substratum is
overlain by a finer textured cover, either occurring to
the surface or up to the plaggen epipedon. This cover
contains similar amounts of clay and silt and has the
same median sand size in the soils of Terraces I and II,
but in the soils of Terrace III it has a distinctly
different particle-size distribution, notably with a
much coarser sand fraction, less clay and silt in the
subsoil, and a much coarser sand substratum. Thus,
Terrace III can be separated from Terraces I and II on
particle-size analyses, but the last two can hardly be
separated on this basis. This conclusion confirms
that of van den Broek and Maarleveld (1963).
    The colours of the subsoils found on the various
terraces are indicated in Table 1. A characteristic
colour seems to be 7.5YR 5/6 on all the terraces except
IV, but on each terrace the subsoil colours may be somewhat redder or yellower. The colour of the reddish bands
with 7-8% clay in the sand substratum almost invariably
is 7.5YR 5/6. The quoted colours also pertain to major
or minor parts of the groundmass in the imperfectly or
poorly drained soils.
    The colour separation of Terraces I, II and III
suggested by van den Broek and Maarleveld (1963)
(5YR colours on Terrace I, 7.5YR colours on Terrace
11 and 10YR colours on Terrace III) is not confirmed
by Table 1.

Micromorphology of Illuviation Features

(1) Quantitative aspects:
    The amounts of illuvial clay determined by point
count in the soils is given in Table 1. They are
expressed as a profile-index in %cm (Miedema and
Slager, 1972). The profile-indices are generally in
the moderately high class but differ from soil to soil,
indicating a variable amount of total illuvial clay.
It is clear from the data that there are no systematic
differences between the terraces.
    Table 1 also shows that the thickness of the Bt

R. Miedema et al.

horizon is not uniform and ranges from only 60 cm in
the imperfectly drained Grubbenvorst soil to 125 cm in
the well drained Lomm 3 and the imperfectly drained
Roamweg soil. The thickness of Bt horizons,therefore,
does not show any systematic difference between the
terraces nor between present drainage conditions.
    These data agree very well with those for soils in
Late-Weichselian Rhine deposits (Miedema et al., 1978;
Miedema, 1982) and with the figures for total clay
illuviation in Weichselian loess soils (Miedema and
Slager, 1972). The occurrence of clay illuviation
phenomena in present day imperfectly drained soils has
already been described by Miedema et al. (1978) for
soils on Late-Weichselian Rhine terraces, which are
similar in age to the Meuse soils. This suggests a
characteristic degree of clay illuviation related to
Late-Weichselian soil-forming conditions. This concept,
introduced by Hoeksema and Edelman (1960), was confirmed
in case studies of clay illuviation in Late-Weichselian
deposits in the Netherlands previously published by
our group in co-operation with others (Bouma et al.,
1968; Kowalinski et al., 1972; Van Oosten et al., 1974;
Miedema and Slager, 1972; Miedema et al., 1978;
Miedema, 1982; Van Schuylenborgh et al., 1970; Slager
et al., 1976, 1978). In contrast , Holocene Meuse and
Rhine deposits of various ages do not show clay illuviation features (Van den Broek and Maarleveld 1963;
De Bakker, 1965; Miedema, 1982). In a recent publication
on well drained Rhine soils of various ages, Schröder
(1979) showed that clay illuviation is present in
Late-Weichselian and Pre-Boreal soils, but is virtually
absent in the younger soils.
    The texture differentiation in these soils is
predominantly a matter of sedimentation as previously
described, and the pedogenetic clay illuviation may or may
not be superimposed on these differences. Therefore,
conclusions based upon textural differences without
thorough micromorphological study are hazardous,
especially in fluvial soils. The largest amounts of
micromorphologically detectable illuvial clay occur in
the sandier subsoils.

(2)   Qualitative aspects:
      In well drained positions, irrespective of terrace
level, illuvial clay bodies include papules and quasi-
cutans which have largely been biologically translocated (Fig. 2a and b). This agrees with results
previously published (Miedema and Slager, 1972).
Furthermore the majority of these illuviation features
are subject to brunification, the cutans and papules
being uniformly impregnated with iron oxides, thereby
losing part of their bright yellow birefringence
(Fig 2c and d).

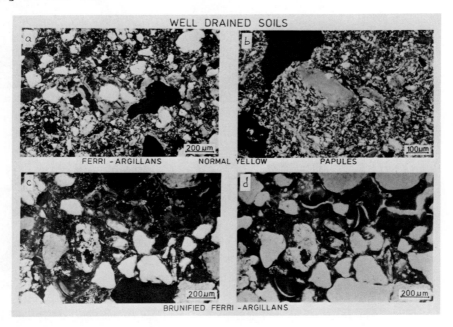

Fig. 2.

526   R. Miedema et al.

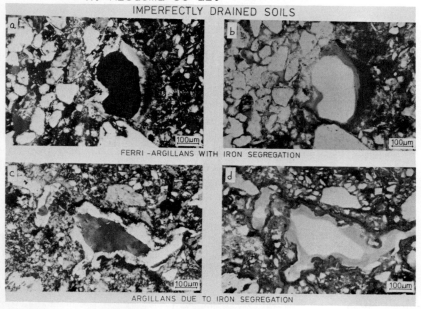

Fig. 3

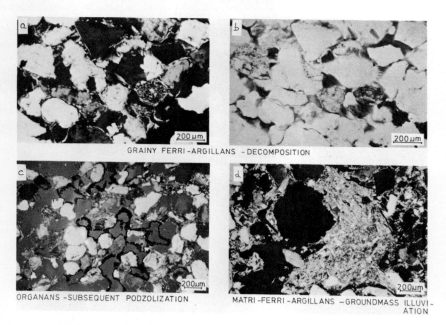

Fig. 4

In soils now imperfectly or poorly drained, clay
illuviation took place under past conditions of better
drainage. In the brown parts of the groundmass some
yellow and brunified cutans remain, but the majority
have been modified to pale yellow or white features
that have lost part of their original iron content in
combination with partly iron/manganese covered
features due to iron segregations elsewhere in the groundmass (Fig. 3a - d). In these greyish parts some cutans
have a characteristic grainy appearance due to clay
decomposition (Brinkman et al., 1973; Van Oosten et al.,
1974) (Fig. 4a and b). In the sandier variants of these
present day hydromorphic soils a subsequent hydromorphic
podzolisation is indicated by organans and ferrans
covering clay illuviation features (Fig. 4c).
      The illuviation bodies in soils on Terraces I, II
and III have various morphological features that can be
grouped and interpreted to help understand the modification
of clay illuviation features by subsequent soil-forming
processes, which are mainly related to drainage position,
but independent of the relative age of the terraces.
      In both well drained soils and the upper part of
soils now imperfectly or poorly drained, matri-ferri-
argillans (Van Schuylenborgh et al., 1970) occur to a
limited extent due to groundmass illuviation as a
result of agricultural activities (Fig. 4d). These
matri-ferri-argillans normally cover clay illuviation
features where the two occur together. A few cutans
formed in situ due to mineral weathering, as described
by Mermut and Pape (1970, 1973), have also been observed
in both well drained and imperfectly to poorly drained
soils.

CONCLUSIONS
      The distinguished terrace levels can only be separated
unambiguously on the basis of elevation. Particle-size
distribution allows differentiation of Terrace III from
I and II, and clay illuviation phenomena provide a means
of separating Terrace IV from the others. Subsoil
colour and a quantification of clay illuviation phenomena
cannot be used to separate Terraces I, II and III.
      Conclusions based on textural differences within
the profiles are unreliable, as they are not entirely due

to clay illuviation. Differences in present drainage
condition are clearly reflected by changes in the
morphology of the clay illuviation features on all three
levels due to subsequent soil-forming processes.

ACKNOWLEDGEMENTS

We wish to acknowledge particularly Mr. Jeronimus
who very skilfully prepared the illustrations for the
poster, Mr. Van Druuten for part of the photographic
work, and Mrs. Beemster and Mrs. Bouter for typing
the manuscript.

REFERENCES
Bouma, J., Pons, L.J. and Van Schuylenborgh, J. 1968.
     On soil genesis in temperate humid climate.VI.
     The formation of a glossudalf in loess (silt loam).
     Neth. J. Agric. Sci., 16, 58-70.
Brinkman, R., Jongmans, A.G., Miedema, R. and Maaskant,
     P. 1973. Clay decomposition in seasonally wet,
     acid soils: Micromorphological, chemical and
     mineralogical evidence from individual argillans.
     Geoderma, 10, 259-270.
De Bakker, H. 1965. Tonverlagerung in Fluszablagerungen
     verschiedener Art. Mitteilungen der Deutschen
     Bodenkundlichen Gesellschaft, Bd. 4, S., 123-128.
Hoeksema, K.J. and Edelman, C.H. 1960. The role of
     biological homogenisation in the formation and
     transformation of gray-brown podzolic soils.
     Trans. 7th Int. Congr. Soil Sci., Madison, Vol.
     IV, 402-405.
Kowalinski, St., Pons, L.J. and Slager, S. 1972. Micro-
     morphological comparison of three soils derived
     from loess in different climatic regions. Geoderma,
     7, 141-158.
Mermut, A. and Pape, Th. 1970. Micromorphology of two
     soils from Turkey with special reference to in-
     situ formation of clay cutans. Geoderma, 5, 271-
     281.

Mermut, A. and Pape, Th. 1973. Mikromorphologie von in situ gebildeten Tonhaütchen in Böden. Leitz Mitteilungen fur Wissenschaft in Technik, Bd. V, 8, S, 243-246.

Miedema, R. 1982. Soil formation, microstructure and physical properties of Late-Weichselian and Holocene Rhine deposits in the Netherlands.Ph.D. Thesis (in preparation).

Miedema, R. and Slager, S. 1972. Micromorphological quantification of clay illuviation. J. Soil Sci., 23, 309-314.

Miedema, R., Van Engelen, E. and Pape, Th. 1978. Micromorphology of a toposequence of Late Pleis-tocene fluviatile soils in the Eastern part of the Netherlands. In: M. Delgado, (Ed), Micromorfologia de Suelos. Universidad de Granada, Vol.1, 469-501.

Schröder, D. 1979. Bodenentwicklung in Spätpleistozänen und Holozänen Hochflutlehmen des Niederrheines. Habilitationsschrift, Bonn, 296 Seiten.

Slager, S., Jongmans, A.G. and Pons, L.J. 1976. Fossil and recent soil formation in Late Pleistocene Sand deposits in the Eastern part of the Netherlands. Neth. J. Agric. Sci. 24, 173-178.

Slager, S., Jongmans, A.G., Miedema, R. and Pons, L.J. 1978. Fossil and recent soil formation in Late Pleistocene loess deposits in the Southern part of the Netherlands. Neth. J. Agric. Sci., 26, 326-335.

Van den Broek, J.M.M. and Maarleveld, G.C. 1963. The Late-Pleistocene terrace deposits of the Meuse. Mededelingen van de Geologische Stichting, Nieuwe Serie no. 16, 13-25.

Van Oosten, M.F., Slager, S. and Jongmans, A.G. 1974. The morphology and genesis of pseudogley phenomena in a Pleistocene loamy sand in the Netherlands. Neth. J. Agric. Sci., 22, 22-30.

Van Schuylenborgh, J., Slager, S. and Jongmans, A.G. 1970. On soil genesis in temperate humid climate VIII. The formation of a "Udalfic" Eutrochrept. Neth. J. Agric. Sci., 18, 207-214.

TRANSLOCATION OF FINE EARTH IN SOME SOILS FROM AN AREA IN MID WALES

T.R.E. Thompson

Soil Survey of England and Wales,
Rothamsted Experimental Station,
Harpenden,
England

ABSTRACT
     Thin sections from profiles described and sampled
during a soil survey of land around Arddleen (SJ260160) in
Wales reveal coatings of translocated silt and clay
intimately mixed in some horizons with organic matter.
Such coatings are found in all the principal soil types
in the area except young well drained brown alluvial
soils with exceptionally high faunal activity. Even
if formed in such soils, coatings would be short lived.
In ungleyed horizons ped faces are darkened by the
coatings; in gleyed horizons they are light grey. They
are often rippled, this being especially evident at
depth in alluvial gley soils. In thin section, they
are composed of various fractions between fine sand and
fine clay. Some are weakly banded but most are
homogeneous and between crossed polarisers have a
speckled extinction pattern lacking the preferred orientation of argillans. In any one profile the composition
of the coatings is coarsest near the surface, the
proportion of finer material increasing with depth.
     As an example, this paper illustrates and describes
the coatings in a profile of the Denbigh series (typical
brown earth). Comparisons are drawn with similar
features reported elsewhere. Their importance in the
classification of the soils and the implications to
farming are discussed.

LOCATION
     Arddleen is close to the Welsh border with England
and lies between the rivers Vyrnwy and Severn. Annual
rainfall is approximately 800 mm with a mean daily
temperature of 15°C in July and an average maximum

potential cumulative soil moisture deficit of between
100 mm and 140 mm at the beginning of August.
     The area is at the eastern limit of the Lower
Palaeozoic rocks that dominate Welsh geology. They are
represented at Arddleen by soft shelly fine sandstones
and siltstones which, although now decalcified, are
thought to cause the high pH values (>7) of many
subsoils in the district. The rocks are easily broken
down and form the basis of the thin glacial drift which
covers the hills, and the fine silty alluvium which the
rivers Vyrnwy and Severn have deposited on their valley
floors.

SOILS
     The area is dominated by soils from three major soil
groups (Avery, 1980). In drift on the hills and on
terraces, brown earths are developed with stagnogley
soils occupying sites where seepage occurs and/or the
drift is impermeable. The Severn alluvium is subject to
seasonally high groundwater and alluvial gley soils are
extensive. The Vyrnwy valley is narrower and it is here
and on the Severn levee that brown alluvial soils are
developed in alluvium free of groundwater.
     As part of the soil survey of the Arddleen area,
representative profiles of the common soil series were
identified, dug to 150 cm and described using standard
Soil Survey terminology (Hodgson, 1976). Bulk samples
were taken from each horizon for mechanical and chemical
analysis and Kubiena tins of undisturbed soil collected
for impregnation and thin sectioning (Avery and
Bascomb, 1974).
     Coatings of various particle-size distributions are
found in the lower subsoils of profiles belonging to
the brown earths, stagnogley soils and alluvial gley
soils in the Arddleen area. They are absent from most
brown alluvial soils in which faunal activity is intense.
In this case the absence of coatings may simply reflect
the considerable disruption to which the subsoils are
subject.
     The fine texture of the coatings makes them most
easily seen in soils with appreciable medium and coarse
sand fractions. For this reason, profile SJ21/6286

Table 1. Analyses of Denbigh profile, SJ21/6286.

Horizon	A	Bw1	Bw2	BC	2Cu	3Cx
Depth (cm)	0-18	18-41	41-63	63-87	87-98	98-120
Sand (600μm-2mm)%	8	8	12	11	9	18
(200-600μm)%	5	5	9	10	12	10
(60-200μm)%	7	8	8	13	26	12
Silt 2-60μm%	53	54	42	44	44	40
Clay <2μm%	27	25	29	22	9	20
Fine clay <0.2μm%	2	7	5	3	2	4
Fine clay/total clay	0.1	0.3	0.2	0.1	0.2	0.2
$CaCO_3$ equivalent%	0.4	0.1	nil	nil	nil	nil
Organic carbon%	2.9		1.1			0.3
pH in water (1:2.5)	7.2	7.5	7.5	7.5	7.5	7.4
pH in 0.01 M $CaCl_2$ (1:2.5)	6.8	7.0	7.0	7.0	6.9	6.8

T.R.E. Thompson

Table 2. Coatings in Denbigh profile, SJ21/6286.

Horizon designation	Section depth (cm)	Description of coatings
Ap	8	None observed
Bw1	24	None observed
Bw2	46	Many homogeneous coatings of mixed fine silt to fine clay lining and filling voids. They are weakly oriented with speckled extinction between crossed polarisers
BC	56	Many coatings of mixed fine silt to fine clay lining and filling voids. Most are homogeneous, some exhibit banding of alternate clay and $< 20 \mu m$ silt laminae; weakly oriented with few patches of moderately oriented clay (Fig. 1)
2Cu	93	Many clay coatings mostly of $< 0.5 \mu m$ clay, some with 1-1.5 μm occasionally $< 2 \mu m$ clay centres where the voids are filled; moderately oriented but still speckled (Fig. 2)
3Cx	104	Many coatings of fine clay, with coarser clay centres if void is filled; moderately oriented; uneven distribution throughout section

of the Denbigh series, a typical brown earth, is
used to illustrate the coatings which vary little from
one soil type to another.
     The large organic content of the coatings is a most
noticeable feature in upper horizons and only in the
lowest horizons are they confused with argillans. Table 1
gives the analyses for the profile; Table 2 describes
the appearance of the coatings in thin section. All
horizons are porphyroskelic with asepic or weakly sepic
plasmic fabrics. Drawing together these descriptions
and those from other soil profiles the coatings are
characterised by the following properties.
1. Coatings occur on ped faces and line and/or fill
channels and voids. In the field they are sometimes
rippled in appearance.
2. They contain a range of particle sizes from fine
sand to fine clay. They are neither argillans as defined
by Brewer (1964) nor matrans (Bal,1973.) but conform to 'poorly
sorted deposits' in Federoff's (1974) classification of
mechanical accumulations.
3. In ungleyed horizons they are darker than the matrix
and appear to contain organic matter. In gleyed horizons
they are light grey.
4. They can be banded but are normally homogeneous.
5. They have low birefringence increasing to moderate
in coats with more fine clay, and the extinction pattern
between crossed polarisers is speckled.
6. With increasing depth in the profile, there is a
trend to a finer particle-size distribution, i.e. from
fine sand with silt and clay in upper horizons to clay
with or without some fine silt at depth.

DISCUSSION
     Traditionally, because of its climate and mostly
acidic rocks, Wales has not been considered to have soils
with illuvial horizons containing translocated clay. Only
over limestone have soils with argillic horizons been
recognised (Thompson,1978). However, Rudeforth (1970)
reports illuvial humus, clay and silt in a Denbigh series
profile from North Cardiganshire, and Thompson (1978)
records 'illuvial clay and silt' in soils around Holywell
in loamy and clayey glacial drift. The advent of more

extensive micromorphological studies, and of deeper
sampling (to 150 cm), may bring to light further
examples of translocation of this type in England and
Wales.   That pipes in land drainage schemes are blocked
by silt and clay implies that such material is mobile
in a large number of soils.   Under different climatic
regimes Brammer (1971) reports illuvial clay, silt and
humus in naturally and artificially flooded soils in
East Pakistan, Spain and Malaya calling them gleyans.
They are speckled in appearance and have low birefringence similar to those reported from Arddleen but
are only recorded in gleyed horizons.
        An important aspect is the relationship between the
coatings described and argillans particularly in
respect to the classification of the soils in which they
occur.   Brammer (1971) considers gleyans and argillans
to be genetically distinct while noting the difficulty
sometimes encountered in separating the two, particularly
in clayey soils.   Evidence from soil at Arddleen is
inconclusive.   In most profiles possessing clay with
silt coatings there is a range from silt + fine sand
+ clay coatings to, with depth, clay with a small amount
of silt, or entirely clay (Figs. 1 and 2).   It must be
noted, however, that where coatings dominated by fine
clay occur and meet the requirements of argillans,they
are separated from adjoining coarser translocated material
by a clear boundary.   It is possible therefore that the
argillans have been shrouded by more recent clay with
silt coatings and that the two have been formed by
separate processes.   In line with the USDA soil
classification (Soil Survey Staff, 1975),Avery (1980)
distinguishes soils with horizons containing amounts
of translocated clay (argillic B horizons) from otherwise
similar soils without such horizons.   The argillic
horizon must however contain 'strongly oriented clay-size
material in coats (argillans) or intrapedal concentrations
(papules or disrupted argillans)'.   Because of this,
none of the soils in the Arddleen area is considered to
have an argillic B horizon as they lack coatings with
strong orientation.   Whether soils containing amounts of
translocated silt and clay warrant separation from
those lacking such material is open to consideration;
at present they are grouped together by Avery (1980).

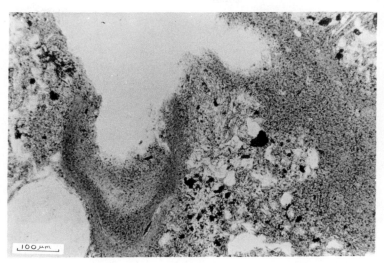

Fig. 1. Clay-with-silt coatings in BC horizon of
         SJ21/6286 (Denbigh series): One coating
         possesses weak banding with thin striae of
         fine clay, the other (right) is homogeneous.
         The speckled appearance is very clear here in
         plain light, and remains so between crossed
         polarisers.

     Brammer (1971) concludes that in East Pakistan,
clay with silt coatings form from topsoil material
dispersed under anaerobic conditions during flooding.
The similarity between his gleyans and the coatings
found in soils at Arddleen is striking,and yet most
of these soils are never flooded and some such as
SJ21/6286 show no signs even of poor profile drainage.
Topsoil puddled and poached by stock or machinery is
a possible source of dispersed material and indeed
results in shortlived anaerobic conditions in the
surface layer. Cultivation could also be instrumental
in the dispersion of soil particles (Greenland,1977).
Greenland's warnings for some soils of possible effects
on subsoil permeability of continuous arable cultivation
due to the translocations of material into deep horizons,
are significant. The relationship between clay with silt

Fig. 2.   Clay and clay-with-silt coatings in the 2Cu
          horizon of SJ21/6286 (Denbigh series).   The
          clay coatings are noticeably less speckled
          (to left of large pore) whereas those with
          considerable coarse clay and some silt are
          similar to the coatings in Fig. 1.   There is
          a thin shroud of speckled coating over most
          of the finer textured clay coatings.

coatings and cultivation is an area for future research.

ACKNOWLEDGEMENTS
        The author wishes to thank Dr. P. Bullock and
C.P. Murphy for the preparation of thin sections and
for much helpful advice during the preparation of this
paper.   C.P. Murphy took the photomicrographs.

REFERENCES
Avery, B.W. 1980.   Soil Classification for England
        and Wales (Higher categories).   Soil Surv. Tech.
        Monogr. 14., Harpenden, England.
Avery, B.W. and Bascomb, C.L. (Ed), 1974.   Soil Survey
        Laboratory Methods.   Soil Surv. Tech. Monogr. 6,
        Harpenden, England.
Bal, L. 1973.   Micromorphological Analysis of Soils.
        Neth. Soil Surv. Paper No. 6.   Neth. Soil Surv. Inst.,
        Wageningen, 174 pp.

Brammer, H. 1971. Coatings in seasonally flooded soils.
    Geoderma, 6, 5-16.
Brewer, R. 1964. Fabric and Mineral Analysis of Soils.
    John Wiley & Sons, New York, 470 pp.
Federoff, N. 1974. Classification of accumulations of
    translocated particles. In: G.K. Rutherford (Ed),
    Soil Microscopy.   The Limestone Press, Kingston,
    Ontario 695-714.
Greenland, D.J. 1977. Soil damage by intensive arable
    cultivation: temporary or permanent? Phil. Trans.
    Roy. Soc., London, B, 281, 193-208.
Hodgson, J.M. 1976. Soil Survey Field Handbook. Soil
    Surv. Tech. Monogr. 5., Harpenden, England.
Rudeforth, C.C. 1970. Soils of North Cardiganshire.
    Memoir of Soil Surv. Great Britain.
Soil Survey Staff. 1975. Soil Taxonomy. A basic system
    of soil classification for making and interpreting
    soil surveys. Agricultural Handbook 436. U.S.D.A.,
    Washington, D.C.
Thompson, T.R.E. 1978. Soils in Clwyd II: Sheet SJ17
    (Holywell). Soil Surv. Record No.50. Harpenden,
    England.

CLAY ILLUVIATION IN CALCAREOUS SOILS

J. Aguilar[1], J.L. Guardiola[2], E. Barahona[2], C. Dorronsoro[3] and F. Santos[3]

[1] Dept. of Pedology, Faculty de Farmacia, Granada, Spain
[2] Station Experimental of Zaidin, C.S.I.C., Granada
[3] Dept. of Pedology, Faculty de Farmacia, Salamanca, Spain

ABSTRACT
    Forty four profiles have been studied of which three have been selected to demonstrate illuviation in calcareous materials.
    In profile development a period of erosion is invoked leading to the removal of A horizons. Following this, there is translocation of clay from the superficial Bt horizon downwards into the calcareous horizons below. Abundant argillans occur to a depth of 2.8 m. Illuviation in calcareous material (27-65% $CaCO_3$) is thought to be due to dispersion of clay from overlying de-calcified horizons particularly after the first rains of the season before the solution becomes calcium-rich.

INTRODUCTION
    In the classical view of clay illuviation, leaching of soluble salts is a necessary precursor to dispersion of the clay which can then be translocated to lower horizons. In the deeper horizons a decrease in electro-kinetic potential is experienced due to the lack of down-ward movement of the water front or to a $CaCO_3$-rich layer, and the clay is deposited. However, ideas are now changing. There have now been a number of observations of clay skins within calcareous materials. Several possible causes of this have been advanced. For example, Reynders (1972) noted that clay translocation is a common process in some calcareous Moroccan soils with argillans forming deep in the calcareous horizons. Dispersed clay and dissolved carbonate are illuviated in some New Mexico soils according to Gile (1970). Some authors (e.g. Allen and Goss, 1974) have invoked the possibility of argillic horizons in decalcified material being re-calcified following a further period of aeolian deposition of

Table 1. Main Analytical Data.

Profile	Hor.	Depth (cm)	Particle-size analysis (%)				C (%)	CaCO3 (%)
			Sand	Silt	Clay	Fine clay		
30	A1	0-13	34.2	44.5	21.1	5.2	1.97	2.2
	B1	13-35	37.0	40.4	22.6	5.6	0.53	2.9
	B2t	35-70	10.3	42.2	47.5	15.3	0.29	1.3
	B3ca	70-120	28.0	43.0	28.9	5.8	0.24	45.8
	C1ca	>120	31.3	31.2	37.2	5.6	0.10	46.1
36	Ap	0-20	44.0	43.3	12.7	3.3	1.34	
	IIB21t	20-45	26.9	36.5	36.6	20.3	0.51	
	IIB22t	45-65	18.9	37.8	43.4	22.6	0.34	
	IIB23t	65-88	11.9	22.8	65.5	27.8	0.23	
	IIB3tca	88-116	22.8	28.9	48.6	18.9	0.30	26.6
	IIC1ca	116-160	30.4	44.8	24.8	5.5	0.24	61.9
	IIC2ca	>160	16.4	67.8	15.7	6.0	0.28	64.6
40	Ap	0-25	29.5	52.9	17.9	4.7	1.43	3.2
	B2	25-40	27.2	44.6	28.3	6.8	1.07	5.4
	B3ca	40-63	28.8	39.3	32.0	5.4	0.39	27.5
	C1ca	63-100	13.7	47.3	39.4	3.2	0.53	34.9
	C2ca	100-130	14.4	37.3	48.0	11.4	0.36	23.5
	C3ca	130-155	8.9	39.9	50.9	10.4	0.31	11.4
	IIB2tb	>155	7.8	49.6	42.7	11.3	0.33	1.4

Table 1:   Main Analytical Data   (Contd.)

Hor.	COLE (fine earth)	COLE (whole soil)	pH	Exchangeable cations (me/100g)				CEC (me/100g)	
				Na	K	Ca	Mg	NaAc	NH$_4$Ac
A1	0.022	0.009	7.4	0.4	1.4	22.5	6.4	14.3	16.9
B1	0.008	0.007	7.6	0.1	0.4	7.9	1.9	10.5	31.5
B2t	0.066	0.066	6.9	1.2	1.2	29.9	7.7	27.3	31.5
B3ca	0.036	0.032	7.8	0.8	0.3	46.0	3.9	22.4	
C1ca	0.021	0.018	7.8	2.7	0.5	44.2	7.8	22.4	
Ap	0.035	0.028	7.1	0.1	1.1	9.7	0.7		10.4
IIB21t	0.069	0.067	5.4	0.5	1.2	23.7	3.1		25.8
IIB22t	0.074	0.073	6.5	0.6	0.6	28.1	2.6		32.9
IIB23t	0.080	0.078	6.6	0.6	0.6	36.3	2.5		34.0
IIB3tca	0.083	0.061	7.1	0.7	0.5	53.0	1.8	29.6	
IIC1ca	0.018	0.013	7.3	0.7	0.4	42.8	0.9	19.1	
IIC2ca	0.035	0.032	7.5	0.5	0.2	38.8	0.7	13.0	
Ap	0.033	0.032	7.6	0.4	1.6	31.9	1.1	26.0	
B2	0.052	0.050	7.5	0.4	1.2	43.6	1.1	26.4	
B3ca	0.033	0.026	7.6	0.5	0.7	48.0	1.2	23.1	
C1ca	0.035	0.030	7.4	0.5	0.6	47.7	0.9	24.7	
C2ca	0.061	0.052	7.3	1.0	0.6	59.5	1.7	30.4	
C3ca	0.086	0.085	7.2	0.5	0.7	57.9	1.9	33.2	
IIB2tb	0.064	0.061	7.0	0.3	0.7	25.1	1.4	29.2	

calcareous dust. Thus, although clay skins are clearly
evident within calcareous horizons of some soils, the
exact conditions of clay translocation are often not
clearly established.
       Forty four profiles have been investigated with
particular reference to the role of clay translocation
in Spanish soils. Three profiles,representing particularly
soils in which there is clay illuviation within calcareous
horizons,are reported on here.

RESULTS
       The main analytical data for the three profiles are
given in Table 1. The micromorphological descriptions
of the 3 profiles with special reference to clay translocation and presence of carbonate are as follows:

Profile 30
       A1   Contains features similar to a former B horizon.
       B1   No illuviation; no carbonates.
       B2t   Many argillans and ferri-argillans around grains,
                 pedological features are dispersed in the
                 matrix; no carbonates.
       B3ca   Very many argillans, some coating planar voids,
                 others with varying degrees of disruption and
                 integration in the matrix; some very well
                 developed argillans coating calcareous material;
                 abundant carbonates forming micritic and
                 sparitic nodules; disrupted argillans within
                 micritic nodules, less in sparitic nodules.
       C1ca   As above but with fewer argillans.

Profile 36
       Ap   Degrading illuviation papules in an other-
                 wise apparently present day Ap horizon; no
                 carbonates.
       IIB21t   Argillans and ferri-argillans surround nodules
                 and grains and, more or less modified by
                 stress, coat planar voids; some disruption of
                 argillans; ferruginous and manganiferous
                 nodules; no carbonates.
       IIB22t   Very clayey matrix in which is embedded numer-
                 ous illuviation argillans; no carbonates.
       IIB23t   Rare argillans and ferri-argillans around

skeleton grains; most illuvial clay
(less than above) present as papules within
matrix; carbonate nodules.

IIB3tca   Some argillans around nodules; illuvial
clay well preserved in sparitic nodules,
more disrupted in micritic nodules; clay
and carbonate co-exist in the matrix; many
carbonate nodules.

IIC1ca   Abundant argillans, well preserved against
pores, more disrupted away from pores;
common randomly distributed sparitic and
micritic carbonates; some compound cutans
consisting of calcite against the pore, an
argillan, and another calcite coat
immediately against matrix; common micritic
carbonate nodules, some containing disrupted
argillans.

IIC2ca   Argillans adjoin $CaCO_3$ matrix (Fig. 1a);
zone in which clay appears to invade the
carbonates; away from pore, argillans
become disrupted (Fig. 1b).

IIC4ca   Very abundant illuvial clay papules; argillans, often more or less disrupted (Fig.1c),
in large planar voids and around smaller
voids.

IIC6ca   Abundant void argillans showing little or no
disruption; very abundant illuvial clay
papules in some cracks, 1-3 mm thick; compound cutans consisting of calcite and clay
(Fig.1d); abundant micritic and sparitic
carbonates; some zones in which $CaCO_3$
appears to invade and disrupt illuvial clay.

Profile 40
Ap   Some illuvial clay suggesting an atypical Ap
horizon; some volcanic quartz and carbonates.

B2   Rare argillans and ferri-argillans; few to
common papules; some rounded carbonate
nodules, a few with clayey bodies inside.

B3Ca   Rare disrupted papules; matrix consists of
carbonates and clay; occasional zones of
carbonates without clay.

C1ca   Virtually no illuvial clay; most matrix

consists of approximately equal proportions
of clay and carbonates; few zones in which
one or other is dominant.

C2ca   Rare illuviation; abundant calcareous
          material.

C3ca   Much carbonate; some disrupted papules;
          carbonate deposits on pore walls; abrupt
          boundary with horizon below except in a few
          parts where carbonate appears to invade
          horizon below.

IIB2tb   Buried argillic horizon; strongly disrupted
          clay coatings; no carbonates.

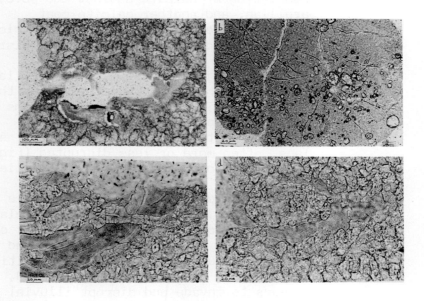

Fig. 1. Photomicrographs from Profile 36: (a) Clay
cutan lining pore in a calcareous matrix, IIC2ca
horizon; (b) Disruption of clay cutans by CaCO$_3$
in a first stage, IIC2ca horizon; (c) Disruption
of clay cutans by CaCO$_3$ in a second stage,
IIC4ca horizon; (d) Alternate deposition of
CaCO$_3$ and clay in a pore, IIC6ca horizon. All
PPL.

DISCUSSION

The main processes involved in the development of Profile 30 are shrink-swell, clay translocation and decalcification - secondary calcification. The B horizon has a skel-mo-vosepic plasmic fabric reflecting the dynamic nature of the soil material under seasonal shrink-swell regimes. The pressures and the micro-churning arising from the process produces preferred orientation around voids, grains, and as patches within the matrix. The degree of disruption is high not only because of shrink-swell processes but also in some horizons apparently through calcite crystal growth. It is of interest that nodules formed of micritic calcite have more features of clay disruption than do those of sparitic calcite.

Clay translocation is a feature of the non-calcareous B2t horizon but also of lower calcareous horizons in which the translocated clay occurs in a more or less disrupted state.

For the formation of this soil, the hypothesis is advanced that following a period of eluviation and illuviation of clay to form a Bt horizon, there was a period of erosion which removed the eluvial horizons. There followed a further period of eluviation - illuviation, this time with eluviation of dispersed clay from the existing Bt horizon and illuviation in the calcareous horizons below.

The Bt horizon of Profile 36 contains both undisturbed (Fig. la) and disrupted argillans (Fig. lb and c). The disruption can be traced to high shrink-swell activity indicated by a high COLE. Some disruption may be associated also with calcite crystal growth; as in Profile 30,micritic calcite appears to have more disruption associated with it than does sparitic calcite. Disruption is generally stronger in upper horizons than lower ones.

The relationship between clay and carbonate appears to be variable and sometimes difficult to unravel. There are zones, e.g. in the IIC2ca horizon, in which clay appears to invade the mass of carbonates whereas the opposite observation is made in some lower horizons. In some horizons, argillans coat carbonate nodules, in others, particularly lower ones, small calcite crystals occur within the illuviation argillans. There are some compound

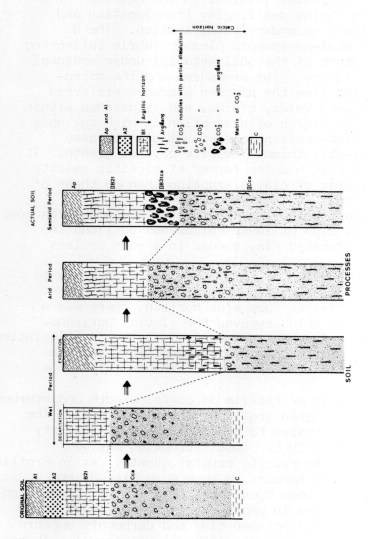

Fig. 2. Possible climatic changes during formation
of Profile 36.

cutans consisting of argillans and calcitans.

Several climatic changes are considered to have taken place in the formation of this profile (Fig. 2). The original soil is thought to have been an Alfisol with A1, A2 and Bt horizons over a Cca horizon.   There followed:

Stage 1.   Moist period;   erosion of the A1 and A2 horizons.

Stage 2.   Moist period;   strong decalcification;   development of thick argillic horizon with argillans forming also in the C horizon;   formation of calcic horizon by carbonate deposition at the base of the B horizon.

Stage 3.   Arid period;   disruption of argillans in former C horizon to produce papules.

Stage 4.   Moist period;   fresh decalcification reducing the lower limit of the argillic horizon;   inactive carbonate nodules coated with argillans.

The climatic changes would correspond to the late Würm (25,000-10,000 years), Hypsithermic (or Altithermic) (9000-2000 years) and the final semi-arid environment to the present day climate.   This scheme fits in with the profile age of 16,000 years calculated by the Arkley method (1963).

The presence of argillans in a calcareous matrix is still difficult to explain but, as Wieder and Yaalon (1978) have pointed out, although carbonate may prevent the dispersion of clay, it does not prevent the movement of already dispersed clay if pores are large enough. A decalcified Bt horizon above a calcareous horizon could be the source of dispersed clay.

Profile 40 shows clearly a buried Bt horizon below a calcic horizon.   The argillic horizon, which is non-calcareous, is strongly disrupted presumably due to high shrink-swell, possibly in combination with seasonal hydromorphism.

REFERENCES

Allen, B.L. and Goss, D.W. 1974. Micromorphology of paleosols from the semi-arid high plains of Texas. In: G.K. Rutherford (Ed), Soil Microscopy. The Limestone Press, Kingston, Ontario, 511-525.

Arkley, R.J. 1963. Calculation of carbonate and water movement in soil from climatic data. Soil Sci., 96, 239-248.

550   J. Aguilar et al.

Gile, L.H. 1970. Soils of the Rio Grande Valley border
     in southern New Mexico. Soil Sci. Soc. Amer.
     Proc., 34, 465-472.
Reynders, J.J. 1972. A study of argillic horizons in
     some soils in Morocco. Geoderma, 8, 267-279.
Wieder, M. and Yaalon, D.H. 1978. Grain cutans resulting
     from clay illuviation in calcareous soil material.
     In: M. Delgado (Ed), Micromorfologia de Suelos.
     Universidad de Granada, 1133-1158.

DIAGNOSTIC PROPERTIES AND MICROFABRICS OF ACID B HORIZONS: COMPARISONS WITH PODZOLIC Bs, CAMBIC Bw AND ACID ELUVIAL HORIZONS

P. Aurousseau

Laboratoire de Science du Sol, 65 Rue de Saint Brieuc, 35042 Rennes, France

ABSTRACT
    A special subtype of cambic B horizon is proposed. The concept for the horizon is defined according to the arrangement of two microfabrics, one characterised by a loose arrangement of void aggregates, 70-100 μm diam., the other by a denser arrangement of primary aggregates. The model colour of the horizon is 7.5 YR or 10YR 5/6. In the field the structure is micro-crumb or subangular blocky breaking to micro-crumb. The main physico-chemical characteristics are pH (water) 4.5, base saturation 15% and exchangeable Al 2-6 me/100 g. The main morphological, microscopic and analytical differences from Bs, E and Bw horizons are detailed.

SEQUENCE OF SOILS
    On schists, sandstones and granites, it is common to describe Cambisols in upslope and Luvisols in downslope positions (Fig. 1a). Depending on substrate, it is possible to distinguish two pedological situations.
    On Pre-Cambrian schist, the cambic horizon in the upslope position has a blocky structure and a moderately high base saturation. This horizon corresponds to a subtype of cambic Bw horizon defined below. In this situation, there is lateral continuity in the sequence between the Bw horizon and the downslope Bt horizon, by gradual increase in the argillic character of the horizon downslope (Fig. 1b).
    On Pre-Cambrian sandstone, the cambic horizon upslope has a micro-crumb structure and low base saturation. It corresponds to a subtype of cambic horizon termed ALE (Aurousseau et al., 1978). Downslope, the leached (lessivé) horizon of the Luvisol has similar character-istics. The Bt horizon of the downslope soil has accumulic rather than argillic characteristics (Fedoroff,

P. Aurousseau

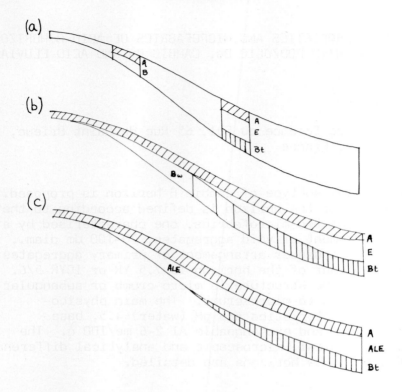

Fig. 1. (a) Generalised horizon sequence with
         Cambisols upslope and Luvisols downslope;
         (b) Sequence on Pre-Cambrian schist;   (c)
         Sequence on Pre-Cambrian sandstone.

1974). In this sequence, lateral continuity occurs
through the intermediary of the subsurface ALE horizon
(Fig. 1c).
     The ALE horizon is widespread in the Armorican Massif.
On quartzitic sandstone it occurs in podzol profiles as a
transitional horizon between the spodic B horizon and the
C horizon (Fig. 2). This ALE horizon is in lateral
continuity with an ALE horizon on the break of slope in
the position of the eluviated (lessivé) horizon.
     These three sequences illustrate the main boundaries

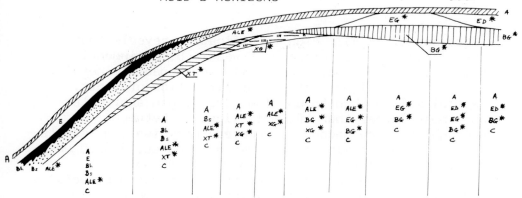

Fig. 2. Horizon sequences on quartzitic sandstone
in the Armorican Massif. (* = notations of
the author).

that need to be defined:
    ALE/Bw
    ALE/Bs
    ALE/E

THE ALE HORIZON
    The ALE horizon corresponds to a subtype of cambic
B that is acid and low in bases.

Horizon Characteristics
    The field criteria for defining the ALE horizon are
colour and structure. The modal colour is 7.5YR or
10YR 5/6. Usually, the hue is 7.5YR or 10YR, occasionally
5YR. The value is generally 5 or 6, occasionally 4
and the chroma 6, occasionally 4.   The structure is
subangular blocky breaking down to micro-crumb. Other
associated characteristics are: large porosity, low
density, very friable consistency, pH 4.5 - 4.9, very
low base saturation (about 15%), 2 -6 me/100 g
exchangeable aluminium, and clay fraction dominated by
vermiculites with hydroxy-aluminous interlayers.

Microscopic Organisation
    The ALE horizon results from the spatial arrangement
of two types of organisation (Roussel, 1980, 1983).
The first (loose) organisation results from the loose
packing of round aggregates, 70 - 100µm diam.

(Aurousseau, 1978). High magnification reveals that the aggregates are composed of weakly birefringent, pale coloured plasma. The easily weatherable minerals, such as micas, lack coatings of alteration products. Particles less than 50 μm diam. are within aggregates and those larger than 50 μm lie loosely between aggregates. The second (connected) organisation results from a denser packing of rounded aggregates, 70–100 μm diam., and skeletal grains larger than 50 μm.

In the upper parts of the ALE horizon and especially where the ALE horizon lies directly below an organic A horizon, the loose organisation is dominant. The denser connected organisation increases downwards towards the base of the ALE horizon and laterally towards horizons affected by hydromorphism.

Bw HORIZON

The Bw horizon corresponds to a subtype of cambic B horizon characterised by moderately acid conditions and moderate base saturation.

Properties of the Bw Horizon

The diagnostic field characteristics are a modal colour of 7.5YR or 10YR 5/4 and a moderately developed blocky structure. Base saturation is about 40% and pH in water between 5.5 and 6.0. The dominant clay mineral is usually vermiculite sensu stricto.

Microscopic Organisation

The Bw horizon is characterised by a continuous organisation, consisting of silt- or sand-sized particles in a fairly abundant, birefringent plasma. Some clay coatings are present. The pores are mainly cavities and channels of undoubted biological origin and fissures which partly define blocky aggregates. At high magnification thin clayey alteration products surround easily weatherable minerals.

The form of the edges of the easily weatherable minerals and the presence of coatings as well as the microscopic organisation indicate a likely relationship between the Bw horizon and Bt horizons. This explains the transition in sequence 2 (Fig. 2) between Bw and Bt horizons.

Bs HORIZON

The Bs horizon described corresponds to a friable spodic horizon in the terminology of De Coninck and Righi (1983).

Properties of the Bs Horizon

The diagnostic field properties are a modal colour of 7.5YR or 10YR 5/8 or 6/8 with a subangular blocky structure breaking down to micro-crumb. This type of horizon is also characterised by: pH in water between 4.7 and 4.9, about 2 me/100g exchangeable Al, pyrophosphate extractable Fe and Al, and clay mineral suite often dominated by vermiculite with hydroxy-aluminous interlayers.

Microscopic Organisation

The Bs horizon results from the spatial arrangement of two types of organisation: (a) A loose organisation of rounded aggregates, 20-30 μm diam., strongly coloured by amorphous iron and diffuse organic matter; particles of less than 20 μm compose the aggregates; particles larger than 20 μm occur loosely between the aggregates. (b) A denser connected organisation of the same elementary aggregates.

The main microscopic differences between Bs and ALE horizons lie in the smaller size of elementary aggregates in the Bs than in the ALE horizon and in the more highly coloured plasma in the Bs. The loose organisation is dominant in the case of Bs horizons compared with the ALE horizon in which the more connected organisation is more extensive.

E HORIZON
Properties of the E Horizon

The field diagnostic properties are a modal colour 10YR 6/3 or 6/4 and a poorly expressed blocky or lamellar structure. The dominant mineral in the clay fraction is often quartz which arises from micro-division or weathering of the silty quartzose skeleton.

Microscopic Organisation
     This consists of a continuous distribution of sand-
or silt-sized quartz grains. There are usually only
small amounts of plasma which has a fairly low
birefringence. The edges of weatherable mineral grains
lack coats of alteration products.

CONCLUSION
     The scheme presented here is not advanced as a
universal scheme for the micromorphological description
of four horizons or subhorizons ALE, Bw, Bs, E. It is
based largely on experience gained in silty or sandy
silt soils of the Atlantic areas of Europe on primary
or Pre-Cambrian foundations. Soil description in terms
of a microscopic organisation enables each horizon to be
related to one or two characteristic microscopic
organisations.

REFERENCES
Aurousseau, P. 1978. Caractérisation micromorphologique
     de l'agrégation et des transferts de particules
     les sols bruns acides. Cas des sols sur granite du
     Morvan. In: M. Delgado (Ed), Micromorfologia de
     Suelos. Universidad de Granada, 655-667.
Aurousseau, P., Curmi, P., Calvez Le Bars, Y. and
     Roussel, F. 1978. Characterisation of deep horizons
     in catenas with podzolic surface development.
     Abstracts. 11th Int. Congr. Soil Sci., Edmonton,
     220-221.
De Coninck, F. and Righi, D. 1983. Podzolisation and the
     spodic horizon. In: P. Bullock and C. P. Murphy
     (Ed), Soil Micromorphology. A.B. Academic Publishers,
     Oxford, 389-417.
Fedoroff, N. 1974. Classification of translocated
     particles. In: G.K. Rutherford (Ed), Soil Micro-
     scopy. The Limestone Press, Kingston, Ontario,
     695-713.
Roussel, F. 1980. Etude d'une toposéquence sur schistes
     pourprés de Montfort. Application aux problèmes de
     mise en valeur forestière sur les sols dégradés
     dans la région de Rennes. Thèse Dr. Ing., Rennes,
     215 pp.

Roussel, F. 1983. Horizons and microscopic organisations
    characteristic of degraded soils on Cambrian
    schists in central Brittany. In: P. Bullock and
    C.P. Murphy (Ed), Soil Micromorphology.
    A.B. Academic Publishers, Oxford, 559-565.

HORIZONS AND MICROSCOPIC ORGANISATIONS CHARACTERISTIC OF DEGRADED SOILS ON CAMBRIAN SCHISTS IN CENTRAL BRITTANY

F. Roussel

Centre Régional de la Proprieté Forestière, 8, Place du Columbier, 35100 Rennes, France

ABSTRACT
      Horizons and micro-organisations characteristic of degraded soils have been studied in a sequence on Cambrian schists in the Armorican Massif. The horizons in the upper part of the sequence are developed in an acid, well-drained, silty material. The horizons in the downslope part of the sequence are developed in a silty to silty clay, acid, reducing material. The main three horizons ED, BD and XD are described in this study.
      The XD horizon, below 90 cm depth, has two domains differing in colour and structure, with well differentiated texture. Under the microscope, there are clearly two different micro-organisations: a light domain with very pale or non-existent plasma and lacking coatings and a rusty domain with very dense plasma and thick ferri-argillans.
      The BD horizon, at 40-90 cm depth, also has two comparable domains but the contrast between the characteristics associated with eluviation and those with illuviation is less.
      The ED horizon, from 10-40 cm, has only one micro-organisation, with clear characteristics associated with eluviation and loss of iron comparable with the light coloured domains of the BD and XD horizons.

INTRODUCTION
      The study of horizons and microscopic organisations characteristic of degraded soils is based on a sequence of soils on Cambrian schists in the Armorican Massif (Fig. 1). The sequence is 50 m long with an average slope of 4%. Its representativeness has been checked over the whole mapping area on Cambrian schists.

F. Roussel

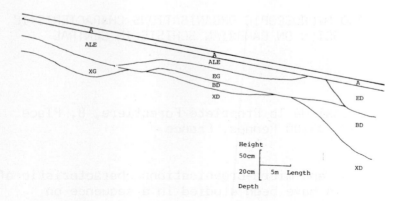

Fig. 1. Horizon sequences in relation to slope.

    In the upslope part of the sequence the horizons are
developed in a well-drained acid silty material. The ALE
horizon as defined by Aurousseau (1976, 1983) and Roussel
(1980) is present in this part of the sequence.
    The horizons in the downslope part of the sequence are
in an acid, reducing,silty to silty clay material.
Horizons described include: ED, BD, XD (Roussel, 1980).
Beneath the organic horizons, the ED horizon occurs from 10
to 40 cm. It is defined by a single type of organisation
termed ED. The BD horizon, which occurs from 40 to
90 cm depth, is defined by two types of organisation -
BD1 and BD2. The XD horizon below 90 cm, is also defined
by two types of organisation, XD1 and XD2.

DESCRIPTIONS OF ED, BD AND XD HORIZONS

XD Horizon
    Two very contrasted colours are noted in the field:
(a) light grey (10 YR6/1) and(b) strong brown (7.5YR5/6).
There is a relationship between the spatial organisation
of these two domains and structure:
- the light grey colour occurs in fissured zones which
  define the lamellar structure.
- the rusty domain constitutes the central part of blocky
  aggregates which are the dominant structural units.
Under the microscope, each of these domains has a
particular organisation:

(a) XD1 organisation: It comprises about
50% of the volume of the horizon. The skeleton is
essentially composed of smooth polyhedral quartz grains
50-80 μm diam. Some muscovite grains without aureoles
of clayey plasma and with sharp edges are present also.
There are clusters of quartz without plasma in some parts
but generally the quartz grains are embedded in a pale
yellow plasma. It is a skelsepic organisation. Coatings
are rare; where they occur they are rather thin, poorly
microbedded and with inclusions of dark granules. There
are also rare orange papules about 100μm diam. containing
dark granules, clearly discrete in the plasma.

(b) XD2 organisation: It comprises about 30%
of the volume of the horizon. The skeleton is comparable
to that in XD1 but is partly masked by a very dense
ferruginous clayey plasma. The organisation is skelsepic.
There are several types of clay accumulation: (a)
stratified coatings, 50-100μm thick, in vesicular pores.
The first layer consists of orange-brown ferruginous clay,
poorly microbedded and weakly birefringent. The second
layer is bright yellow, microbedded, with alternating
beds of clay and silt; (b) bright yellow, poorly microbedded clay coats to fissures; (c) brown and yellowish
brown, continuous, well microbedded coatings in some
vesicles; (d) rare papules, less than 2mm diam.; whose
microbedding is masked by a homogeneous reddish brown
material.
        These complex accumulations of clay-silt-iron show
little or no integration with the matrix. They
characterise an accumulic type of organisation (Fedoroff,
1973). The XD2 organisation occurs in direct contact with
XD1. The transition can be sharp (<0.1mm) or more
gradual (about 1mm).

(c) Supporting analytical results:
In parallel with microscopic examinations, other
selected determinations have been made. There is a
difference of 6% clay between XD1 (16%) and XD2 (22%)
as well as a difference in iron oxide content of 4%
(2.5% in XD1, 6.3% in XD2). Mineralogically, there are
also significant differences between the two organisations:

(i) organisation XD2 (rusty) has mainly kaolinitic and
micaceous clay minerals; goethite and lepidocrocite
are the main iron oxide minerals; 10Å minerals
are commoner than 14Å minerals and, among the latter,
vermiculite sensu stricto, rather than hydroxy-aluminous
vermiculite, is dominant. This mineralogical suite is
similar to the parent material and the minerals themselves
are but little altered; (ii) organisation XD1 (light grey)
compared with XD2 has mainly 14Å rather than 10Å clay
minerals and the former generally have hydroxy-aluminous
interlayers.

        (d) Interpretations: In the XD horizon there
are two contrasting organisations: (a) XD1 with
impoverishment of iron and clear eluviation characteristics and (b) XD2 with iron enrichment and clear illuviation
characteristics.
        The mineralogical differences between the two
organisations can be attributed to degradation of 2:1 clay
minerals which affects the light grey XD1 organisation.
Because of the spatial arrangement of this organisation
within the horizon, degradation influences the zones
of preferential water circulation. In the rusty XD2
organisation, the very abundant iron oxides/hydroxides
can play a part in protecting the clays from degradation.

BD Horizon

        A comprehensive network of oblique fissures 0.2-0.3
mm wide partially defines a coarse polyhedral structure.
Two organisations unrelated to this structure provide
a variegated pattern: (a) BD1 organisation corresponds
in the field to a grey (10YR 7/3) domain and (b) BD2
organisation to a rusty (7.5YR 5/8) domain.

        (a) BD1 organisation: This constitutes about
60% of the volume of the horizon. Two zones arranged
in more or less clear beds are distinguished; (a) a
zone with skeleton of quartz grains about 50μm diam.
without plasma and (b) a zone with skeleton of quartz
grains about 10-20 μm diam. with a plasma that is fairly
open.

The micas lack clayey coats.   Some pores of 0.10-
0.15 mm diam. are lined with a denser optical zone containing iron, with a gradual transition to the matrix.
Few very fine coatings occur in some pores and there
are also some rare fragmented papules.   The fissures are
not coated.

(b) BD2 organisation: The quartzose skeleton of
this organisation has similar heterogeneity to that in
BD1 organisation.   The organisation has a brown to dark
brown plasma and a skelsepic plasmic fabric.   Ferruginised strands lack a homogeneous optical density.   Near
the centre of the strands the plasma becomes more
opaque.   Vacuolar pores, 0.5-1mm diam., are numerous,
and have rather thin (20-50μm) discontinuous light yellow
coatings.   Along the fissures which define the fine
polyhedral structure there are fine light yellow coatings
of iron and clay or dark brown coatings of organic matter
and clay.   There are also some fragmented papules.
Numerous coatings are more or less embedded in the matrix,
typical of an argillic organisation (Fedoroff, 1974).

(c) Supporting analytical results:
On selected samples analysed, clear differences have been
noted between BD1 and BD2 organisations.   There is a
difference of 6% clay between BD1 (20%) and BD2 (26%)
and a difference of more than 3% Fe_2O_3 (BD1 2.2%, BD2
5.3%).   In terms of mineralogy, BD has a mineralogical
suite comparable to that of XD1.   The suite in BD2 is
intermediate between XD1 and XD2.

(d) Interpretation: As in the XD horizon, there
are morphological and analytical differences between
the two types of BD organisation: (a) BD1 has eluviation
characteristics and has lost iron;   (b) BD2 contains both
illuvial clay and iron enrichment.
The contrast is less from a mineralogical point of
view.   This is not necessarily due to less degradation
but may rather reflect differences in soil material
(weathering material in situ or transported material)
in which pedogenesis has taken place.

ED Horizon

This very massive, weakly porous horizon is characterised by a single type of organisation. As in BD1, two zones can be distinguished, disposed in sub-horizontal altered layers, 100-500μm thick: (a) a coarse textured layer, with quartz grains 40-50μm diam., without plasma, and completely discoloured and (b) a fine textured layer composed of quartz and some muscovite grains 10-20μm diam. set in a fairly open light yellow plasma. Micas have a washed appearance.

Independent of bedding there are more or less rounded weathered pieces of schist, 0.5 to 1mm diam., composed of quartz grains 30-50μm diam. set in very dark iron-stained plasma. Around these pieces of schist are brownish red aureoles with a diffuse boundary to the adjacent matrix. The horizon contains 12% clay and 1.5% iron oxides. Its mineral suite is comparable to that of BD1 and XD1 organisations. This horizon has very advanced characteristics of eluviation and removal of iron. Processes of degradation affect the whole of the horizon.

CONTRIBUTIONS OF DEMARCATED HORIZON ORGANISATIONS

Description and understanding of a pedological medium in terms of the sequence-horizon-organisation aids a solution to the following problems:
1. With chemical and mineralogical analyses carried out on individual organisations it is possible to:
   a) establish a parallel between micromorphological observations and analyses.
   b) understand marked analytical variation of some horizons composed of several primary microscopic organisations present in varying proportions in a series of samples from one horizon (Roussel, 1980).
2. Description of horizons in terms of organisations leads to better specification of horizon boundaries.
   From the description of horizons given earlier, there is clearly a morphological gradation between ED and BD1 organisations. The boundary between ED and BD horizons is more clearly defined by the appearance of BD2 organisation. The length of the transition can be defined by measuring the distance between the level at which the BD2 organisation appears and the level at which

the BD organisation is at least 40% by volume.   In this
sequence it is 10cm.   There are other types of boundary
between BD and XD horizons; for example,a morphological
difference between BD2 organisations (argillic type)
and XD2 organisations (accumulic type), and finally,
a difference in the arrangement of the organisations
as a function of the structure of the horizon.   On the
one hand the arrangement of BD1 and BD2 organisations
has no visible relationship to the structure of the BD
horizon whereas the XD1 and XD2 are linked to the
structure of the XD horizon.
      These examples show that the boundaries between
horizons can be defined in several ways: (a) appearance
or disappearance of an organisation;   (b) modification
of a type of organisation;   (c) modifications between
two or more organisations.
      This approach has been used in the description and
analysis of the horizons and organisations in a sequence
relating to the transition between upslope ALE and
downslope ED, BD and XD horizons.   This description has
necessitated the definition of the EG horizon in a
central position.   The latter shows two types of
organisation.   Organisation EG1 is present in the upslope
ALE horizon (Aurousseau, 1976, 1981;   Roussel, 1980);
EG2 has intermediate characteristics between EG1 and ED
organisations.   Characteristics of eluviation and
removal of iron are not well marked.

REFERENCES
Aurousseau, P. 1976.   Morphologie et génèse des sols sur
      granite du Morvan.   Thèse de Doct. Ing.,Rennes.
      177 pp.
Aurousseau, P. 1983.   Diagnostic properties and micro-
      fabrics of acid B horizons.   Comparisons with
      podzolic Bs, cambic Bw and acid eluvial horizons.
      In: P. Bullock and C.P. Murphy (Ed), Soil Micro-
      morphology. AB Academic Publishers, Oxford, 551-557.
Fedoroff, N. 1974.   Classification of translocated
      particles.   In: G.K. Rutherford (Ed), Soil
      Microscopy.   The Limestone Press, Kingston, Ontario,
      695-715.
Roussel, F. 1980.   Etude d'une toposéquence sur schistes
      pourprés de Montford.   Thèse de Doct. Ing., Rennes,
      215 pp.

MICROMORPHOLOGY OF DEEPLY WEATHERED SOILS IN THE TEXAS COASTAL PLAINS

L. P. Wilding[1], M. H. Milford[1] and M. J. Vepraskas[2]

[1] Soil & Crop Sciences Department, Texas A&M University, College Station, Texas 77843.
[2] Soil Science Dept. North Carolina State University, Raleigh, N.C. 27650.

ABSTRACT
    Four soils along a toposequence on the coastal plain in east Texas were analysed micromorphologically to further elucidate pedological features in relation to previously determined soil moisture regimes.   Upland soils have deep sola with evidence of clay translocation extending to depths of 4 to 7 m.   Infilling of vughs, channels and cracks with oriented clay reaches a maximum at 2 to 3 m, making these Bt horizons relatively imperm-eable.   The A horizons contain few argillans or nodules. Poorly drained soils are characterised by infilled cray-fish krotovinas of silty stratified impervious material sparsely penetrated by roots.   Conditions favour lateral downslope interflow of water above the deep, impervious argillic horizons of the upland soils.   Frequent reductive conditions in the upper Bt horizons have resulted in Fe and Mn translocation, some to lower sola of the upland soils and some through lateral interflow to low-land soils where ferri-argillans, ferrans, sesquans, mangans and Fe-Mn concentrations are common plasma features.   Micromorphological evidence suggests that the Bt horizons of the downslope soils formed contemporane-ously with upland soils after hillslope dissection and before the lower topographic reaches became poorly drained.

INTRODUCTION
    The study area is located in the east Texas Coastal Plains (Fig. 1) along a 3% southeastern-facing hillslope about 50 m above sea level. It has a local relief differential of about 15 m (Vepraskas, 1980).   Soils have formed in mid to early Pleistocene fluviatile-

L.P. Wilding et al.

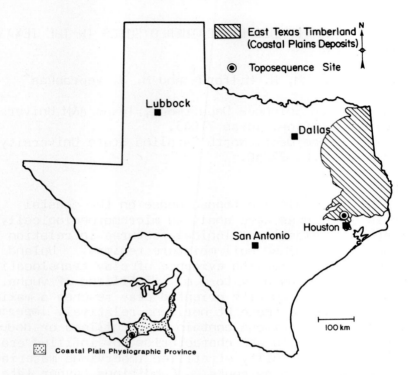

Fig.1.   Location of toposequence site in Texas.

deltaic coastal plains sediments originating during inter-
glacial epochs in response to rising sea levels (Aronow,
1976).   The climate of the area is subtropical, with an
evenly distributed rainfall of about 1200 mm.   The mixed
forest vegetation is dominated by loblolly pine (Pinus
taeda).   Mean annual air temperature is about 21°C.
     The purpose of this paper is to present micromor-
phological evidence to verify the pedological origin of
the deeply weathered soils and to confirm two hypotheses:
1) that Fe nodules are relict features in upper sola;
and 2) crayfish krotovinas contain few macropores and act
to decrease saturated hydraulic conductivity.

METHODS
     Thin sections were prepared and described for
selected horizons following methods and terminology of
Brewer (1976).   Undisturbed oriented clods were impregnated under alternating vacuum and 15-bar pressure with

a Castolite-styrene (3:2) solution. Horizontal and
vertical sections were cut from each clod, then
trimmed, ground and epoxied to glass slides. Each
section was ground to a thickness of about 30 μm.

RESULTS

Morphological, Physical, Chemical and Mineralogical
                    Properties
     Complete characterisation of each of the soils in
the toposequence and observed soil moisture and redox
potentials are available elsewhere (Vepraskas, 1980).
Major morphological features of the toposequence are
illustrated in Fig. 2. Kaolinite is the major clay
mineral with significant quantities of vermiculite-

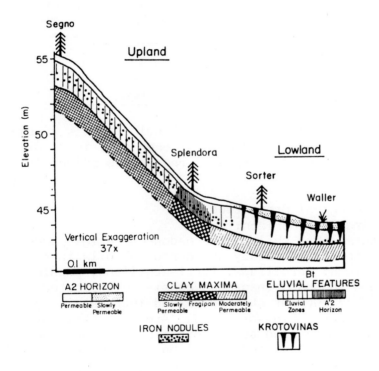

Fig. 2.   Schematic cross section illustrating landscape
          position and major morphological features of
          each soil in the toposequence.

L. P. Wilding et al.

chlorite intergrade and smectite in upper sola. Soil
pHs, base saturation, exchangeable Na percentages (ESP)
and smectite content increase in A2 and Bt horizons
from upper to lower slope components: pHs increase from
5.5 to 7.5; base saturation from 40 to 100%, and ESP
from < 5 to 18%. Clay depth functions are given in
Fig. 3.

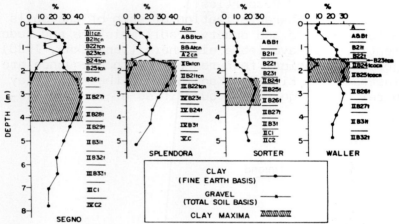

Fig. 3.   Depth functions for total clay (< 2 μm)
          gravel (> 2 mm) for each soil in topo-
          sequence. (Gravel represents primarily iron
          nodules).

Micromorphological Properties
      (a) Argillic horizon:
      Features that reflect the pedogenic origin and land-
scape continuity of deeply weathered subsoils in this
region include: argillans, ferri-argillans, papules,
ferrans and sesquans (neo- and quasi-) distributed along
former and/or current structural and intramatrix con-
ductive elements (Fig.4a and b). Argillic horizons of
clay maxima correspond to microfabrics with maximum
development of illuviated argillans and papules exhibiting
moderate to strong continuous orientation. Up to 75%
of the plasma, as illuviated clay, plugs vughs and
channels, and bridges embedded skeletal grains. Multi-
stage development of these pedological features is
evident from superposition and intersection of compound

cutans that indicate oscillatory oxidising and reducing
conditions on both a macro- and micro-scale. The resulting
slowly permeable, dense porphyroskelic s-matrix with
inundulic plasmic fabric restricts water movement and
favours seasonal perched water-tables of low redox
potentials in superposed horizons.

Evidence that argillic horizons of members lower in
the toposequence developed contemporaneously with upland
equivalents and under better drained hillslope conditions
that predate their silty upper sola include: 1) the micro-
morphic similarity of s-matrix and plasmic fabric
elements; 2) similarity in argillic expression under
markedly different current hydrologic regimes; 3) slowly
permeable silty overburden that restricts infiltration
and eluviation of clay; 4) long periods of saturation
and/or high soil moisture in subsoils that preclude or
minimise translocation of clay to lower depths; and 5)
microfossil, particle-size and elemental indices indicative of a lithologic break between silty overburden and
argillic horizons of lower topsequence members.

(b) Eluvial horizons:
Eluvial horizons have intertextic or porphyroskelic
s-matrices with silasepic plasmic fabrics. Illuvial
features are generally absent; there are a few Fe-Mn
concentrations. Lower chroma structural surfaces and
s-matrix zones in upper Bt horizons are strongly eluviated;
illuviated clays are restricted to higher chroma s-matrix
areas. Micromorphological observations support transitional eluvial to illuvial horizonation of 0.5 to 1 m
thickness in upper sola.

(c) Iron nodules:
Iron nodules are a major morphological feature
concentrated in illuvial horizons of upper slope soil
components (Fig. 2); they rarely extend below 2 m.
Nodules contain skeletal grains in the same proportion
as the s-matrix and often contain embedded grain ferri-
argillans indicative of in situ nodule formation. Where
nodules are prominent, fluctuating seasonal water-tables
produce conditions favouring alternating redox potentials
of iron reduction and oxidation (Vepraskas, 1980). Nodules
are sparse in clearly eluvial horizons. Those in the

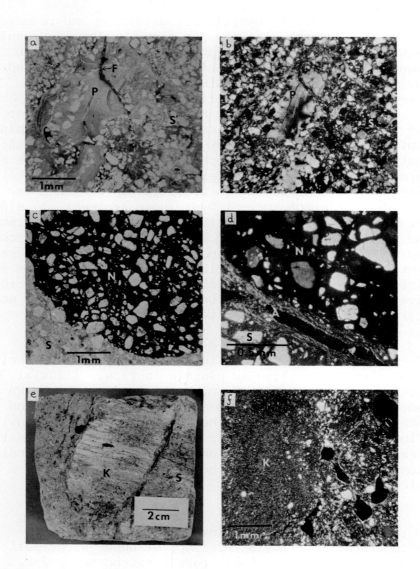

Fig. 4.

eluvial-illuvial transitional horizons have clear
boundaries with many exposed skeletal grains on the
perimeter (1.4 to 3.2 exposed grains/mm of perimeter),
possibly indicating the nodules are degrading (Fig. 4c).
In lower, more illuvial horizons, nodules are often
coated by argillans and have gradual boundaries to the
s-matrix (Fig. 4d); skeletal grains on the perimeter
tend to have a lower frequency of exposure (1.0 to 1.5
grains/mm perimeter) and often are coated by thin ferrans.
This may indicate nodule stability and growth.   Quasi-
and neo- Fe-Mn plasma concentrations, ferrans, mangans and
sesquans occur at depths up to 6 m.

   (d) Krotovinas (pedotubules):
   Silty upper sola of lower toposequence members
are markedly altered by krotovina infillings that comprise
up to 75% or more of the volume of A, A&B and B&A horizons.
These features are readily recognized by banded arcuate
fabrics, strong separation with host material, uniform
sorting of skeleton grains, and granular s- matrices
encompassed by porphyroskelic host fabric (Fig. 4e and f).

Fig.4.   (a,b) Illuviated clay papule (P) subsequently
         transected by a channel ferran (F) in the s-
         matrix (S) of Splendora B21t (200-240 cm) argillic
         horizon. PPL (a) and XPL (b). Note strong band
         extinction of papule and inundulic plasmic
         fabric;   (c) Iron nodule (N) in s-matrix (S) of
         Splendora A'2 (118-143 cm) horizon illustrating
         protruding skeleton grains from the degrading
         nodule surface. PPL;   (d) Same nodule (N) where
         it abuts a high chroma illuvial zone in the
         s-matrix (S) and where nodule contains a ferri-
         argillan (A) at its surface. XPL;   (e) Hand
       specimen of krotovina (K) encased in s-matrix
         (S) of the Waller A&B horizon (13-23 cm);   (f)
         Photomicrograph of krotovina (K) and s-matrix
         (S) in Waller B21t horizon (78-125 cm). XPL.

Krotovinas are very slowly permeable; they contain few
macropores and are texturally banded. Dense massive host
material with non-interconnecting meta-vughs is equally
restrictive; ponded conditions may persist at the
surface of these soils for several months in the
presence of unsaturated A2 and subjacent Bt horizons.

CONCLUSIONS
     Microfabric analysis confirms the pedological origin
of deeply weathered soils of the Texas Coastal Plains.
They have laterally co-extensive argillic horizons that
perch water to form reductive horizons from which clay,
Fe and Mn have been eluviated. Argillic horizons of more
poorly drained lower topographic reaches are contemporaneous with upland equivalents. They must necessarily
have developed under better drained hillslope conditions
and predate their associated silty upper sola.

REFERENCES
Aronow, S. 1976. Geology. In: L. P. Wilding and
     J.F. Mills (Ed), Morphology, Classification and Use
     of Selected Soils in Harris County, Texas. Tour
     Guide, Soil Sci. Soc. Amer. Div. S-5. Texas Agric.
     Exp. Sta., Dept. Soil and Crop Sci. Tech. Report
     76-48, 52 pp.
Brewer, R. 1976. Fabric and Mineral Analysis of Soils.
     Robert E. Krieger Publishing Company, Huntington,
     New York. 482 pp.
Vepraskas, M.H. 1980. Soil morphology and moisture regimes
     along a hillslope in the Texas Coastal Plain. Ph.D.
     Dissertation, Texas A&M Univ., College Station, Texas.

SOME PROPERTIES OF DEGRADED ARGILLANS FROM A2 HORIZONS
OF SOLODIC PLANOSOLS

H.J.M. Morrás

Departamento de Suelos, INTA Castelar, Argentina

ABSTRACT
    Illuviation cutans in A2 horizons of certain solodic
planosols have an altered appearance which differentiates
them from cutans in B horizons. This degradation is
expressed in the form of abundant micro-fissures and
cavity pores which, in extreme cases, give a spongy
appearance to the cutanic plasma. The microanalytical
and mineralogical data indicate that this porosity arises
from the straightforward process of remobilisation and
translocation of clay particles rather than from chemical
weathering.

INTRODUCTION
    During a micromorphological study of solodic planosols
(FAO-UNESCO,1974) in the 'Bajos Submeridionales', Argentina,
clay coatings were noted in the A2 horizons which had a
degraded appearance, differentiating them clearly from
cutans in B horizons (Morrás, 1978).
    Various mechanisms involving alteration and trans-
location have been proposed by different authors to explain
the genesis of these soils (Morrás, 1979). Although
features associated with temporary hydromorphism in these
soils have been noted in this section,the degraded clay
coatings of A2 horizons have rarely been mentioned or
studied in detail.
    One exception is the work of Brinkman et al. (1973)
on pseudogley soils, in which analyses of cutans appear
to indicate decomposition of the clay with release
followed by washing out of components and re-precipitation
of the silica as quartz.
    The cutanic deposits of A2 horizons reported here
show particular characteristics suggesting a different
pedogenetic mechanism.

RESULTS
    Ferri-argillans occur mainly in the upper part of the

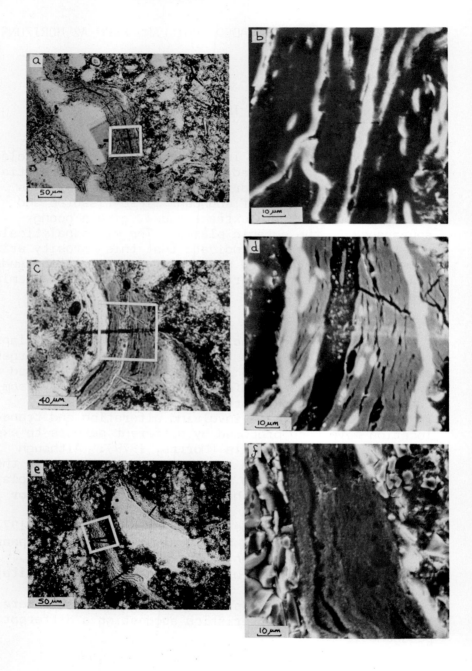

Fig. 1.

Fig. 1. Optical and electron microscopy of cutans:
(a) Argillan from B horizon. PPL; (b) Detail
of (a) by SEM. Between shrinkage cracks (bright
edges), the clayey plasma is homogeneous; (c)
Ferri-argillan from A2 horizon. Note strong
cracking and filling of voids by unoriented
clay. PPL; (d) Detail of (c) by SEM. The
structure degradation of this cutan is evident;
(e) Cutan from A2 horizon composed of an
orange central part and yellowish outer part.
Note the very marked cracking of other cutan.
PPL; (f) Secondary electron image of cutan in
(e) showing high porosity. The spongy nature
appears more marked in the outer part (right)
than in the inner part.

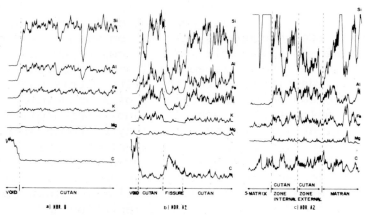

Fig. 2. Microprobe analysis of cutans along traverses
indicated on the respective SEM images in Fig. 1.

B horizon. Siltans and partcutans (compound cutans consisting of siltans alternating with clayey deposits) are
also present, but rare, and at the base of the B horizon
there are hydromorphic illuviation argillans. The normal
argillan is the most frequent type of cutan in the B
horizon. These are usually moderately thick (30-120 μm),
yellow, fine textured, homogeneous and with a high degree
of orientation and birefringence (Fig. 1a).

The larger cutans often contain some micro-fissures corresponding to shrinkage cracks, for the most part parallel with the edge of the void. Under the optical microscope, these fissures appear as dark lines whereas under the electron microscope they appear as bright streaks. Between these fissures the cutanic plasma is homogeneous and without any sign of degradation (Fig. 1b).

Microanalysis of this type of coating across a traverse (Fig. 2a) shows that the chemical composition of it is also uniform, discontinuities only occurring at the shrinkage cracks. Using peak heights and peak ratios as a basis for determining the relative contents of some components the cutan appears to be composed of alumino-silicates in which potassium is slightly more abundant than magnesium. This agrees with the mineralogical data, which showed a predominance of illite and interstratified illite-smectite (Morrás et al., 1981). The amounts of iron and its constant distribution is in accord with its homogeneous association with clay minerals.

In albic and planosolic A2 horizons of these solodised soils, there are some illuviation cutans which are clearly identifiable from the silty matrix poor in plasma. Whereas in the A2 horizons of less developed solodised solonetz, the argillans and ferri-argillans have approximately the same morphology as those of the underlying B horizon, in the more developed albic horizons, the cutans show several forms of degradation and in this respect differ from those in the B. For example, in some ferri-argillans, yellow in plain light, fine textured and showing strong orientation and birefringence in polarised light, there is strong fissuration suggesting some modification of the coating. The coarser fissures are filled with unoriented material of silty clay texture. Within the central part of the voids are unorientated clay deposits showing patchy extinction (left part of Fig. 1c).

SEM shows that in addition to the major fissures there are also numerous small fissures and cavities (Fig. 1d). Unlike cutans from the B horizon, voids in these cutans from the A2 horizon are irregular and thus unassociated with shrinkage cracks. Rather, they express a process of alteration of a cutanic plasma. Judging from a transect microanalysis, the chemical composition

of the cutan does not vary very much. However, the
carbon peak rises above the base level at regular
points across the cutan and there is a corresponding
decrease in other elements at the same points. The
carbon is considered to be derived from the resin,
and the peaks to indicate the presence of numerous
resin-filled pores in the cutan (Fig. 2b).
        A second type of cutan that occurs in the A2 horizons
is shown in Fig. le . The left hand side of this cutan,
in contact with the matrix,is orange, suggesting a
ferri-argillan. The outer zone at the edge of the void,
on which matrix material has been deposited (matran),
has, by contrast, the usual pale colour of an argillan.
The coating contains several large fissures giving it
an altered appearance. Under SEM, the coating has a
spongy appearance because, in addition to the major
fissures observed under the optical microscope, there
are a large number of small cavity-like pores. The
secondary electron image obtained by SEM shows a
difference between the inner ferri-argillan and the
outer argillan in that the former is much less porous
and more homogeneous.
        The microprobe analyses (Fig. 2c) show that the
contents of silica, aluminium, iron and potassium in
the inner zone are slightly greater than those in the
outer zone, whereas the reverse is true in the case of
carbon.
        These results indicate that the difference in
colour and in appearance in polarised light between
the ferri-argillan and the argillan are not due to
significant differences in content of iron, but to the
porosity of the cutanic plasma. There is no evidence
for a chemical or mineralogical alteration of the cutan
from the results obtained.

CONCLUSIONS
        Illuviation argillans in the well developed A2
horizons of some solodic planosols show, in addition to
a sharp contact with adjacent matrix material, an altered
appearance which differentiates them from argillans
in the B horizon. This differentiation manifests
itself in various ways:
        Cutans in some A2 horizons are characterised by

abundant micro-fissures and in some cases by large
numbers of smaller cavity pores giving an overall spongy
appearance. Despite these evident structural differences,
there is no evidence from micro-analyses for the
mineralogical weathering of these clay deposits as was
noted in similar features in other soils (Brinkman et al.,
1973). From these results, it is reasonable to conclude,
supported also by clay mineralogy studies (Morrás et al.,
in press), that the porosity of the cutanic deposits
originates from a process of remobilisation, resuspension
and translocation of clay particles.

      Because of their presence in present day eluvial
horizons, and on the basis of the structural degradation
that has affected them, these cutans can be considered
as relict features. The processes involved lead to a
deepening of the albic horizon at the expense of the upper
part of the B horizon by a mechanism termed secondary
illuviation.

      There is no evidence for removal of iron from the
cutans prior to remobilisation of the clay as has been
noted in other soils subject to surface-water gleying
(De Coninck et al., 1976). Finally, it is interesting
to note that the optical differences between two parts
of the same coat shown in Fig. 1e and f are not linked
to variations in iron content, but rather to differences
in structure.

ACKNOWLEDGEMENTS

      The author is grateful to Professor G. Bocquier
(Université de Paris VII) for making available to him
the electron microprobe of the Pedology Laboratory.

REFERENCES

Brinkman, R., Jongmans, A., Miedema, R. and Maaskant, P.
      1973. Clay decomposition in seasonally wet, acid
      soils: Micromorphological, chemical and mineralogical
      evidence from individual argillans. Geoderma, 10,
      259-270.
De Coninck, F., Favrot, J., Tavernier, R. and Jamagne, M.
      1976. Dégradation dans les sols lessivés hydro-
      morphes sur matériaux argilosableux. Exemple des
      sols de la nappe détritique Bourbonaisse (France).
      Pédologie, 26, 105-151.

FAO-UNESCO.   1974. Soil map of the world.   Vol.1, Legend.
     UNESCO, Paris, 59 pp.
Morrás, H. 1978.   Contribution à la connaissance pédolo-
     gique des 'Bajos Submeridionales' (Province de Santa
     Fe, Argentine).   Influence de l'environnement sur la
     formation et l'évolution des sols halomorphes.
     Thèse Dr. Ir., Université de Paris VII, 184 pp.
Morrás, H. 1979.   Quelques élements de discussion sur
     les mécanismes de pédogenèse des planosols et
     d'autres sols apparentés.   Sci. du Sol, 1, 57-66.
Morrás, H., Robert, M. and Bocquier, J. (in press).
     Caractérisation minéralogique de certains sols
     salsodiques et planosoliques du "Chaco Deprimido".
     Cah. ORSTOM, Sér. Pédol.

ANTHROPIC INFLUENCE ON A MEDITERRANEAN BROWN SOIL

J.L. de Olmedo Pujol

Centro de Edafologia y Biologia Applicade del Cuarto,
Cortijo de Cuarto (Bellavista)
Apartado de Correos 1052,
Sevilla, Spain

ABSTRACT
        A comparison of terraced and non-terraced soils on
schist in southern Spain shows that terracing has
produced a loose, open soil that contains more coarse
material, more evidence of infilling of voids in lower
parts and less carbonate than in non-terraced situations.
The establishment of terracing in this area seems to
produce soils particularly prone to leaching.

INTRODUCTION
        Terraces have been constructed in an area of
Mediterranean brown earths on schist near Malaga in
southern Spain in order to make the best possible
agricultural use of the soils. It is important to know
the influence of management on this type of soil. A
project has been undertaken to obtain this information.
This paper summarises some of the early results of the
investigation using two profiles, one terraced, the
other natural, as examples.

THE SOILS
Natural soil (without terracing)
            Profile I. La Mayora, Malaga, Spain, Parcel 2.

Horizon	Depth (cm)	Description
Ap	0-45	Moist; very dark greyish brown (10 YR 3/2) sandy loam; loose crumb structure; friable to very friable; permeable to very permeable.
C1	45-60	Moist; dark reddish brown (5 YR 3/2) sandy loam; subangular blocky structure; friable to very friable;

Horizon	Depth (cm)	Description
		permeable to very permeable.
C2	60-100	Moist; dark reddish brown (5 YR 3/2) loamy sand; loose granular structure; very permeable.

Terraced soil

Profile II. La Mayora, Malaga, Spain, Parcel 16.

Horizon	Depth (cm)	Description
Ap1	0-40	Moist; very dark greyish brown (10 YR 3/2) sandy loam; crumb to granular structure; very friable; very permeable.
Ap2	40-65	Moist; dark brown (10 YR 3/3) sandy loam; granular structure; friable to very friable; permeable to very permeable.

ANALYTICAL DATA
     The terraced soil contains more organic matter, more clay, more stones > 2 mm, and has a large cation exchange capacity, presumably associated with the higher organic matter content, than the natural unterraced soil (Tables 1 and 2).

MICROMORPHOLOGY
Profile I
     Ap (28-36 cm). Agglomeroplasmic to porphyroskelic related distribution; compound packing voids in most parts, some areas of simple packing voids; plant remains in various stages of decomposition, occurring both discretely and embedded within the matrix; occasional isotropic and anisotropic Fe-rich clay concentrations as grain and void argillans; some clay papules; few discrete calcite grains and some embedded in matrix; in-skel-vosepic plasmic fabric.

Table 1.   Chemical analyses.

Horizon	OM (%)	C (%)	N (%)	C/N	pH	CaCO$_3$ (%)	CEC (me/100g)	P$_2$O$_5$	K$_2$O	Ca	Mg
								(mg/100g)			
Profile I											
Ap	1.86	1.08	0.13	8.3	7.4	0.8	14.5	44	22	31	31
C1	0.62	0.36	0.05	7.2	7.6	0.8	13.3	11	4	36	41
C2	0.19	0.11	0.01	11	7.6	1.4	5.6	3	3	7	23
Profile II											
Ap1	3.0	1.76	0.16	11	7.3	0.4	19.0	24	18	36	45
Ap2	0.9	0.52	0.08	6.5	7.4	0.8	16.0	16	4	17	44

The larger organic matter content and amounts of Ca and Mg are attributable to additions of manure given to the terraced soil to improve its stability and fertility.

Table 2. Physical analyses.

Horizon	Depth cm	>2mm (%)	Coarse sand (%)	Fine sand (%)	Silt (%)	Clay (%)	Moisture equiv.	Permeability
Profile I								
Ap	0-45	25	29.1	36.8	20.8	12.7	17.9	exc.
C1	45-60	22	42.2	22.4	20.8	13.9	15.4	"
C2	>60	30	67.2	13.3	9.8	7.6	6.5	"
Profile II								
Ap1	0-40	52	45.6	13.8	20.1	18.5	19.1	"
Ap2	40-65	58	33.1	29.4	20.6	15.8	16.1	"

The large amounts of coarse particles (>2mm) particularly in
the terraced soil, together with the open structure, means
that the soils are excessively drained.

Profile II
      Ap1-Ap2 (38-46 cm). Agglomeroplasmic and porphyro-
skelic related distributions; compound and simple
packing voids, few craze planes; many platy shale
particles; compound Fe-rich argillans; reddish and
brownish isotropic Fe-rich clay papules; yellowish brown
plasma; some calcite; in-vo-skelsepic plasmic fabric.
      Ap2 (52-58 cm). Agglomeroplasmic and porphyroskelic
related distribution; compound and simple packing voids;
some void infillings of silt and clay; plant remains
as in Profile 1; few pockets of faecal pellets; abundant
shale fragments > 2 mm; some reddish plasma concentrations; some pale, isotropic nodules; more carbonate
nodules than above; skelsepic plasmic fabric.

CONCLUSIONS
      There are a number of features that point to a
considerably more pronounced leaching in the terraced
soil than in the natural unterraced soil. Thus the
terraced soil contains a larger amount of coarse material,
more evidence of infilling of voids in the lower parts
with matrix material and less carbonate, this having
clear dissolution edges.
      The natural soil has stronger pedality and more
expression of edaphic processes leading to a complex
plasmic fabric.
      The current investigation will be extended to
more sites and profiles. These preliminary findings
suggest, however, that the process of terracing aimed
at improving productivity has some disadvantages, in
particular that it creates a loose, open soil which
is prone to strong leaching and rapid downward loss
of material.

MICROSCOPIC INDICATIONS OF PHYSICAL PEDOGENETIC PROCESSES IN ACID SILTY TO LOAMY SOILS OF BELGIUM

E. Van Ranst, F. De Coninck and R. Langohr

Laboratory of Pedology, Geological Institute, State University of Ghent, Krijgslaan 271, 9000 Ghent, Belgium

ABSTRACT
     The nature of pedogenetic processes (clay migration, biological activity and swelling-shrinking) in acid silty to loamy soils of Belgium is related to the mineralogy of the parent material. The soils from the Ardennes have a cambic horizon, sometimes overlying an horizon with weak clay accumulations. The soils developed in the loess of central Belgium have a well developed argillic horizon. X-ray diffraction and chemical analysis show that the former have more weatherable primary minerals (tri-octahedral chlorite and micas), whereas the calcareous loess contains more smectite. The formation of the argillic horizon in the loess soils is due predominantly to translocation and accumulation of these inherited smectites. In the Ardennes, the weathering of chlorite and micas in the top horizons releases large amounts of Fe and Al which have a strong stabilising effect on the micropeds formed by faunal activity. This stability is a feature of the cambic horizon.

INTRODUCTION
     In Belgium, two regions have dominantly silty to loamy soils. Region I is the Massif of the Ardennes (altitude > 350 m and mostly 500-690 m; rainfall 1000-1400 mm/year) having soils with a cambic horizon developed either in a solifluction layer or in situ in the weathering loam of the underlying bedrock. The bedrock is mainly composed of fine-textured sandstones, shales or phyllites of Paleozoic age. The soils, sols bruns acides (Belgian legend: Tavernier and Maréchal, 1962), meet the criteria for Dystrochrepts and Fragiochrepts (Soil Survey Staff, 1975) or Dystric Cambisols (FAO-UNESCO, 1970). Region II is the loess belt of central Belgium (altitude 50-350 m; rainfall

700-800 mm/year) having soils with an argillic horizon :
Hapludalfs and Glossudalfs (Soil Survey Staff, 1975) or
Luvisols and Podzoluvisols (FAO-UNESCO, 1970).

     Under forest, the Inceptisols of the Ardennes and
the Alfisols of central Belgium show striking differences
in morphology. (1) At the top of the profile, the
Inceptisols have, under an ochric or umbric epipedon,
a thick cambic horizon with a stable crumb structure
and many uniformly distributed roots (De Coninck et al.,
1979). The Alfisols have an ochric epipedon and an
A2 horizon with sometimes a platy structure and with
fewer roots distributed irregularly. (2) At greater
depth, the cambic horizon grades into an horizon with
a weakly pronounced angular blocky structure, with a low
porosity, fewer roots, and some features of clay
illuviation (Langohr and Van Vliet, 1979). In the
Alfisols, the A2 overlies a strongly pronounced argillic
horizon with coarse prismatic structure. In the
Glossudalfs, bleached tongues penetrate into the Bt
horizon. The roots present in this horizon are
situated along the prism faces or in the bleached tongues
in the degraded profiles (Van Vliet and Langohr, 1981).

     The presence of a cambic horizon in the Ardennes
rather than an argillic horizon as in central Belgium
has been explained by the higher rainfall in the former
area (Tavernier, 1964). In most years, the rainfall
exceeds the evapo-transpiration during the whole year
and the presence of only a weak structure and the lack
of clay migration would be a consequence of the nearly
permanent moist status of the soils. However Langohr
and Van Vliet (1979) have shown the presence of argillans
in the field as well as in thin sections of well to
moderately well drained soils of the Ardennes. They
concluded that the absence of clay migration in some
sols bruns acides'is not related to the climate but
rather to the parent material.

MATERIALS AND METHODS
     In order to study some pedogenetic processes in
both groups of soils, a number of representative profiles
under forest were sampled. Physico-chemical and mineralogical analyses were made on bulk samples. The clay

content was determined by sedimentation after H_2O_2
pre-treatment and dispersion with sodium hexametaphosphate.
The clay was separated by dispersion with sodium carbonate.
Organic carbon (OC) was determined by the method of Walkley
and Black (Black, 1965). The pH was measured in H_2O and 1N
KCl, in a ratio 1:1. Free iron (Fe_2O_3 dith.) was
determined colorimetrically with orthophenantroline, after
reduction with Na dithionite added in powder to a Na
citrate-$NaHCO_3$ solution at pH 7.3 at a temperature of $75^{o}C$
(De Coninck and Herbillon, 1969). Free aluminium (Al_2O_3
dith.) was determined by atomic absorption. X-ray
determinations were on samples parallel-oriented on glass
slides and dried after removal of free iron oxides. The
 < 2 μm fraction was examined in Mg^{++}- and K^+- saturated
form. The Mg^{++} slides were also diffracted after glycol
solvation for 24 hours in vacuum and the K^+ slides were
diffracted after heating respectively at 350 and $550^{o}C$
for two hours. The diffraction was carried out with a
Philips X-ray apparatus (PW 1050/25) with CoK_{α} radiation.
Thin sections were prepared by impregnating undisturbed
oriented samples with polystyrene UCEFLEX R66 SL diluted
with monostyrene in the presence of cyclohexanone
peroxide and cobalt-octaoate. For the description, the
terminology of Brewer (1964) was used.

RESULTS
     The analytical and mineralogical results of only
two representative profiles are given : one from the
Ardennes with complete absence of clay migration and
one from the loess belt with a well developed Bt
horizon. The micromorphological data also take into
account the transitional situations.

Physico-chemical and Chemical Analyses (Table 1)
     The upper horizons of the soils are acid and
desaturated. The high pH of the C2ca horizon of the
profile in loess is due to the presence of $CaCO_3$.
Organic carbon gradually decreases with depth in the
soils of the Ardennes, while in the loess soils it
shows a clear break between the Al and the subsurface
horizons. There is a clear accumulation of clay
in the Bt horizon of the profile in loess, whereas such

E. Van Ranst et al.

Profile	Hor.	Depth (cm)	Clay %	OC %	pH 1:1		Dith. 0-2 mm		Dith. 0-2 μm	
					H_2O	KCl	Fe_2O_3	Al_2O_3	Fe_2O_3	Al_2O_3
ARDENNES	A1	0-5	12.0	15.1	3.8	3.2	2.76	1.07	13.15	4.94
Soil in	B2	5-25	12.6	4.1	4.4	4.2	3.39	1.84	13.99	6.76
weathering loam	B3	25-43	11.2	1.4	4.6	4.2	2.26	0.99	10.16	3.47
	C1	43-53	12.3	0.6	4.9	4.4	1.84	0.76	7.26	3.66
	C2	53-100	9.2	0.3	5.0	4.4	1.29	0.71	5.52	3.21
	C3	100-120	12.9	0.6	5.4	4.4	1.42	0.54	5.94	2.40
CENTRAL BELGIUM	A1	0-5	15.6	5.6	3.7	3.2	1.19	0.36	6.09	2.28
Soil in loess	A2	5-22	12.0	0.8	4.3	3.7	1.08	0.34	5.51	2.16
	B2t	22-90	19.6	0.2	4.8	3.9	1.46	0.50	5.41	2.25
	B3	90-172	16.2	0.2	5.4	4.1	1.19	0.25	5.24	1.18
	C1	172-192	14.0	0.1	5.4	4.2	1.06	0.19	5.04	1.20
	C2ca	192+	12.6	0.1	7.9	6.9	1.02	0.21	4.43	0.92

Table 1 : Analyses of the two representative profiles

an accumulation is absent in the Ardennes. The amounts of Fe_2O_3 and Al_2O_3, extracted with dithionite in the < 2 mm and the < 2 μm fraction, are much larger in the soils of the Ardennes. This indicates that more weatherable minerals are present in the soils of the Ardennes.

Mineralogy (Figs. 1 and 2)
     The quantitative mineralogical composition of the < 2 μm fraction with chlorite, mica, kaolinite, quartz, feldspars, smectite and interstratified minerals is similar in both groups of profiles, but the relative intensities of the respective minerals suggest large differences in amounts of the minerals.
     The mineralogical composition of the clay fraction of the soil in the weathering loam is dominated by weatherable primary minerals, trioctahedral chlorite and mica, as indicated by strong 001 reflections in the diffractograms of the parent rock. Both the micaceous and chlorite minerals show a gradual transformation toward the top of the profiles : (1) clear broadening and bridging of the 10 and 14 Å spacings, (2) rather common medium reflections at more than 20 Å and between 10 and 14 Å.
     In the loess profile, the Bt, C horizon and the calcareous loess are dominated by swelling minerals : intense reflections at 17 to 18 Å with Mg^{++} + glycol. In the A horizons the 17-18 Å reflection decreases,

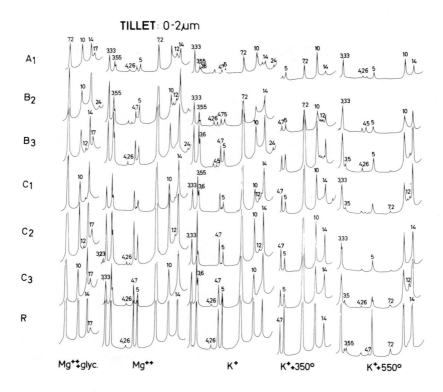

Fig. 1. X-ray diffraction patterns of the < 2 μm
fraction of the profile TILLET in weathering
loam (Ardennes) after dithionite treatment
and saturation with Mg^{++}, Mg^{++} + glycol, K^+,
and after heating of the K^+ saturated sample
to 350° and 550°C.

whereas reflections for the quartz and feldspar become
stronger. These features suggest a preferential migration
of smectite, and a relative accumulation of quartz and
feldspars in the A horizon.

Micromorphology

From the pedogenetic point of view, the following
micromorphological features are important.

The O horizons are dominated by recognisable,
mostly birefringent, strongly comminuted plant remains

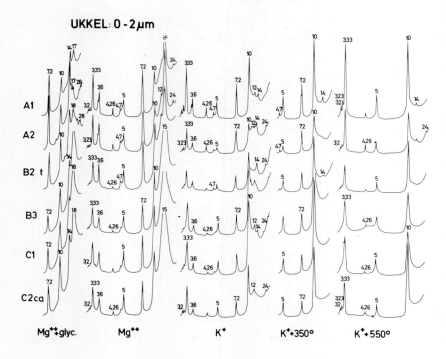

Fig. 2. X-ray diffraction patterns of the < 2 μm
          fraction of the profile UKKEL in loess (Central
          Belgium) after dithionite treatment and
          saturation with Mg^{++}, Mg^{++} + glycol, K$^+$,
          and after heating of the K$^+$ saturated sample
          to 350° and 550°C.

and by faecal pellets and large faunal excrements,
forming loose aggregates with rounded edges (bioformations).
      In the top horizons of the soils of the Ardennes
most of the soil material is in the form of units
similar in shape to the bioformations of the O horizons,
but in which the amount of mineral material gradually
increases downward, although the porosity remains
high (Fig. 3a). Most of these units are part of more
or less clearly recognisable pedotubules and clusters
(Fig. 3b). The s-matrix of the aggregates is asepic.

The upper part of the loess soils is characterised
by a sharp boundary between organic and mineral horizons
(Fig.3c). In the latter, bioformations with rounded
edges are negligible and are observed only where roots
are present. The porosity of the A horizons, especially
the A2, is rather low. The fabric is dominantly
granular (Fig. 3d).

In the B horizons of the soils in the Ardennes the
number of discrete units gradually decreases with a
concomitant decrease in porosity. In the transitional
profiles, the clay plasma becomes locally sepic and
clay concentrations appear. Isolated papules in the
s-matrix increase gradually in number and size
downward and grade into yellow and reddish yellow
argillans, with strong continuous orientation on the
walls of the voids. At some places the clay plasma
becomes masepic.
The Bt horizons of the soils in loess are
dominantly characterised by the presence of yellowish
brown to dark brown ferri-argillans and complex cutans.
In some degraded profiles, greyish argillans are
observed at the bottom of the tongues. The disturbance
of these pedological features decreases with depth.
The plasmic fabric is dominantly sepic.

The C horizons of the soils of the Ardennes are
characterised by an increasing number of lithorelicts
from the underlying substratum.
In the C2ca horizon, or calcareous loess,
microcrystalline calcite is present in the matrix
as well as calcitans lining pores.

DISCUSSION
Indications of two kinds of physical processes
with an opposing effect, and both important for
pedogenesis, can be observed in the thin sections.
The first tends to concentrate clay and silt
particles and is indicated by the presence of argillans,
siltans and complex clay and silt cutans. The
accumulation of clay particles expressed by the presence
of argillans, is important as it brings about a

E. Van Ranst et al.

textual differentiation. The argillans have a low
porosity and a strong layering which is probably due
to a slow deposition of layer silicates out of a
dispersed phase (Van Ranst et al., 1980). Features
arising from this process are clearly dominant in the
soils in loess. In the Ardennes, this process is not
so clearly expressed or is negligible (Fig. 4a and b).

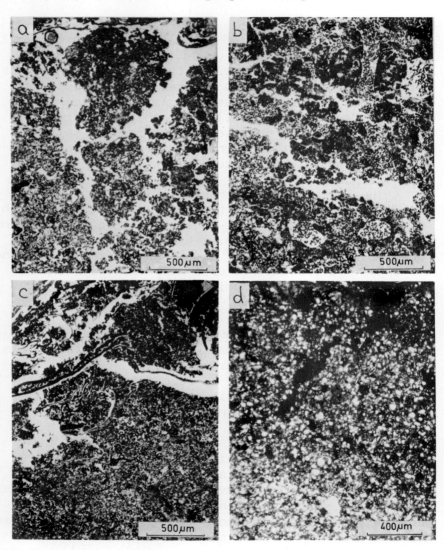

Fig. 3.

The difference in intensity of this process between
both groups of soils is explained by the quantitative
difference in the mineralogical composition of the clay
fraction.   The clay fraction of the soils in loess is
dominated by smectite minerals and in the Ardennes by
chlorite and micas;   the former group migrates more
easily than the latter.
     The second kind of process tends to homogenize
the soil by mixing the different components.   Two
processes are important here:
(1) Features indicating a biological activity have
a much greater importance in the soils of the Ardennes
than in the loess soils.   Most striking in this respect
is the difference in macrostructure and porosity :
stable crumb in the top horizons of the Ardennes, only
weak crumb, and certainly less porous, in the loess
soils.   This corresponds to a microstructure composed of
rounded asepic bioformations containing organic and
mineral material.   The crumbs of the first group have
higher porosity (Fig. 4c and d).
(2) Features associated with shrinking and swelling,
processes indicated by shining faces to the angular blocky
peds, and an absence of argillans.
In thin sections, such features include deformed or
partially transformed argillans, papules (Fig. 5a and
b) and zones with a sepic plasmic fabric (Fig. 5c and d).

Fig. 3.   (a) Upper part (O and A1 horizons) of
          a soil in weathering loam (Ardennes).   Strongly
          comminuted plant remains, faecal pellets
          mixed with mineral particles forming loose
          aggregates with rounded edges and high
          porosity. PPL;   (b) A horizon of a soil in
          weathering loam (Ardennes).   Strongly developed
          crumb structure.   Bioformations present in
          more or less clearly recognisable pedotubules
          and clusters. PPL;   (c) Upper part (O and
          A1 horizons) of a soil in loess (Central
          Belgium).   Fresh and transformed plant fragments
          overlying an A horizon in which bioformations
          are absent and with low porosity.   PPL;
          (d)   A2 horizon of a soil in loess (Central
          Belgium).   Granular fabric with low porosity.
          XPL.

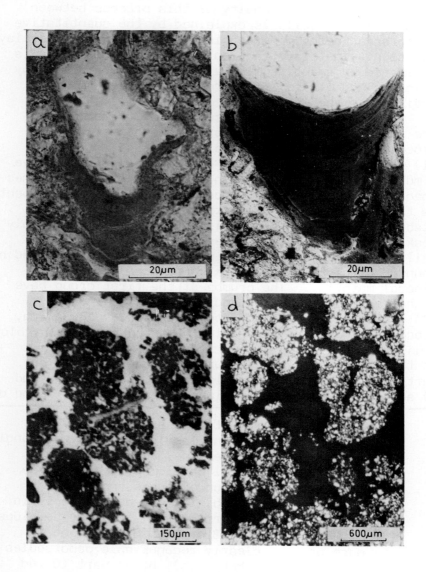

Fig. 4.

The intensities of these two homogenizing
processes are clearly different in both groups of soils.
Features due to biological activity are dominant in the
soils of the Ardennes, but are rather limited in the
loess soils. Features associated with clay accumulation
and swelling-shrinking processes are still dominant in
the deeper horizons of the profiles in the Ardennes
and throughout the whole Bt horizon in the loess profiles.
        Illuviation features, with exception of some
bleached argillans in the bottom of the bleached tongues,
seem to be relict in both groups of soils. Particularly
in the Ardennes, the present-day environment seems not
to favour clay migration because of the release of high
amounts of Fe and Al by weathering of the minerals in the
top horizons. The dominant processes tend to destroy the
differentiations brought about previously. This is shown
by the many bioformations which include papules or
dislocated argillans.   Processes dominant nowadays are
conducive to the formation of the cambic horizon in these
soils.

Fig. 4. (a) B horizon of a soil in weathering
        loam (Ardennes). Argillan in a void. PPL;
        (b) Bt horizon of a soil in loess (Central
        Belgium). Argillan in a void. PPL; (c)
        A horizon of a soil in weathering loam
        (Ardennes). Pedotubule and large bioformations
        with rounded edges and a plant remain, high
        porosity. PPL; (d) B horizon of a soil in
        weathering loam (Ardennes). Bioformations
        showing uniform mixing of mineral and organic
        material and a high porosity. XPL.

E. Van Ranst et al.

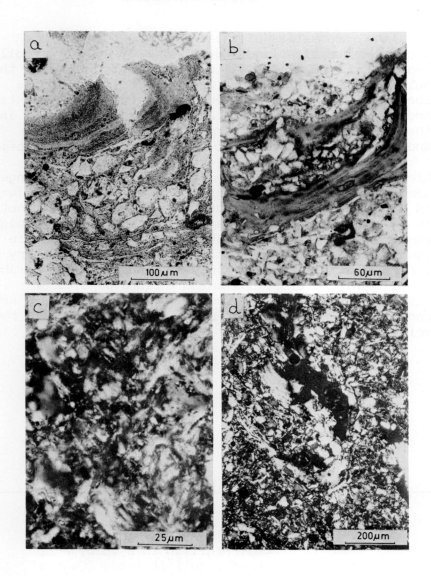

Fig. 5.

Nevertheless, the microfabric and the high porosity of these acid desaturated soils are not typical of cambic horizons developed in the absence of carbonates as defined in Soil Taxonomy (Soil Survey Staff, 1975). Normally they should have a microfabric with little pore space in the matrix and a preferred orientation of particles on the ped faces, due to pressure.

The dominance of bioformations in the soils of the Ardennes can be explained by the larger amounts of free Fe and Al. Indeed, Toutain (1974) and Duchaufour (1977) have shown that free Fe strongly influences the nature of the humus profile : its presence tends to give rise to the formation of an acid mull. Still more important is the influence of free Fe and Al on the stability of the microped formed by faunal activity. They bind material and organic matter (Nguyen Kha, 1973) and have a strong flocculating influence on the clay and organic particles by compressing the double layer. In this way the larger amounts of free Fe and Al are important in the formation and the conservation of the microped,

---

Fig. 5. (a) B horizon of a soil in weathering loam
       (Ardennes). Complex cutan composed of a
       succession of argillans and siltans. PPL;   (b)
       Bt horizon of a soil in loess (Central Belgium).
       Complex cutan composed of a succession of
       argillans and siltans. PPL;   (c) B horizon
       of a soil in weathering loam (Ardennes). Deformed
       argillans and s-matrix with masepic plasmic
       fabric in which the oriented domains are
       remnants of argillans. Low porosity. XPL;
       (d) Bt horizon of a soil in loess (Central
       Belgium). Argillans which are partially incor-
       porated in the s-matrix that has locally a
       sepic plasmic fabric. Low porosity.   XPL.

which are a specific property of the cambic horizon in
these soils.

CONCLUSIONS

1.   The mineralogical composition of the clay fraction
indicates large differences between the two groups of
soils : (a) the soils in the weathering loam of the
Ardennes contain large amounts of weatherable primary
minerals, particularly chlorite and micas. These
minerals undergo a gradual transformation near the top
of the profiles, releasing large amounts of Fe and Al;
(b) the soils in loess are dominated by smectitic
minerals. The decrease in intensity of the smectite
reflections in the A horizons suggests a preferential
migration, resulting in a relative accumulation of
quartz and feldspars.
2.   Two kinds of physical processes can be observed in
thin sections : (a) clay migration tending to a textural
differentiation. This process, indicated by the
presence of argillans, is clearly dominant in the loess
soils, but is less expressed, or in many places
negligible, in the Ardennes; (b) biological activity
and swelling-shrinking processes tending to homogenize
the soil material and to destroy differentiations
brought about by the first process. The biological
activity, especially forming bioformations with high
inter- and intraporosity, is much more important in the
soils of the Ardennes than in the loess soils.
3.   The mineralogical composition seems to be one of
the most important factors governing the physical
pedogenetic processes in these acid silty to loamy soils.
The dominance in the loess soils of smectitic minerals,
which have a strong tendency to migrate, facilitate
the formation of a well developed argillic horizon.
The larger amounts of free Fe and Al, related to
larger amounts of weatherable primary minerals, chlorite
and micas, together with a higher biological activity
forming a more stable crumb structure, are important
for the formation of the typical cambic horizon of the
soils of the Ardennes.

ACKNOWLEDGEMENT
     Contribution to the Research Project "De
mineralogische evolutie in de Bodems van België" (The
mineralogical evolution in the soils of Belgium).
Fonds voor Kollektief Fundamenteel Onderzoek, Belgium.

REFERENCES
Black, C.A. 1965. Methods of Soil Analysis. Amer. Soc.
     Agron., Madison, U.S.A. No.9, 1367-1378.
Brewer, R. 1964. Fabric and Mineral Analysis of Soils.
     Wiley & Sons, New York, 470 pp.
De Coninck, F. and Herbillon, A. 1969. Evolution
     minéralogique et chimique des fractions argileuses
     dans des alfisols et des spodosols de la Campine.
     Pédologie, 19, 159-272.
De Coninck, F., Van Ranst, E., Springer, M.E., Tavernier,
     R. and Pahaut, P. 1979. Mineralogy and formation
     of some soils of the Belgian Ardennes. Pédologie,
     29, 25-69.
Duchaufour, Ph. 1977. Pédogenèse et classification.
     Masson, Paris.
FAO-UNESCO. 1970. Key to the soil units of the Soil
     Map of the World. FAO, Rome.
Langohr, R. and Van Vliet, B. 1979. Clay migration in
     well to moderately well drained acid brown
     soils of the Belgian Ardennes. Morphology and clay
     content determination. Pédologie, 29, 367-385.
Nguyen Kha. 1973. Recherches sur l'évolution des sols
     à texture argileuse en conditions tempérées et
     tropicales. Thèse Doct. Etat Univ. Nancy I.
Soil Survey Staff, 1975. Soil Taxonomy. U.S.D.A.
     Agric. Handb. No.436, Soil Conservation Service,
     Washington D.C.
Tavernier, R. 1964. La genèse des sols de la Belgique.
     In : Soils of Southeastern Europe, Int. Symp. Soil
     Sci., Sofia, June 1963, 79-93.
Tavernier, R. and Maréchal, R. 1962. Soil survey and
     soil classification in Belgium. Int. Soc. Soil
     Sci., New Zealand, 3-12.

Toutain, F. 1974. Etude écologique de l'humification
    dans les hêtraies acidophiles. Thèse Doct. Etat
    Univ. Nancy I.
Van Ranst, E., Righi, D., De Coninck, F., Robin, A.M.
    and Jamagne, M. 1980. Morphology, composition
    and genesis of argillans and organans in soils.
    J. Microscopy, 120, 353-361.
Van Vliet, B. and Langohr, R. 1981. Correlation between
    fragipans and permafrost with special reference
    to Weichsel silty deposits in Belgium and
    Northern France. Catena,8, 137-154.

THE ROLE OF SILICA IN THE FORMATION OF COMPACT AND
INDURATED HORIZONS IN THE SOILS OF SOUTH WALES

E.M. Bridges and P.A. Bull

University College of Swansea and Christ Church College,
Oxford

ABSTRACT
     Soils with compact or indurated horizons are common
in Europe and North America. There is wide support for
compaction by periglacial processes but considerable
divergence of opinion regarding the cementing agents.
Clay, iron, aluminium and organic matter have all been
proposed as indurating materials but do not appear to
be critical in these south Wales soils. Examination by
SEM revealed amorphous silica to be responsible.

     Soils with a high bulk density and brittle fracture
characterising their lower horizons have been described
from many parts of Western Europe and North America.
Similar physical characteristics occur in soils throughout
the northern and western British Isles in both upland and
lowland situations upon a wide range of parent materials
and in many soil groups. In Wales, soils with compact
lower horizons have been identified by Stewart (1961) in
the Aberystwyth district, Wright (1980) in the Llandeilo
district, Crampton (1965) in the uplands of the Brecon
Beacons, Clayden and Evans (1974) in the south Wales
coalfield and by Bridges and Clayden (1971) and Bull and
Bridges (1978) on the coastal lowlands of the Gower
peninsula. These compact horizons are not found in any
fixed relationship to the present soil surface and cannot,
therefore, have developed as a result of contemporary soil-
forming processes. Locally they can be at a constant
depth from the surface as was found over approximately
3 km^2 of soils derived from till and solifluction
deposits from Devonian (Old Red Sandstone) rocks around
Glas Fynydd forest (P. Ashworth, pers. comm.).
     Subsurface horizons or layers with compact brittle

material have been called fragipans (Carlisle et al.,
1957), fragipan horizons (Hodgson, 1974; Matthews, 1976),
indurated horizons or layers (FitzPatrick, 1956; Romans,
1962, 1976; Crampton, 1965) and isons (FitzPatrick,
1971, 1976, 1980). The Soil Survey of England and Wales
(Hodgson, 1974) and the USDA (Soil Survey Staff, 1975)
have defined fragipan horizons as being loamy-textured
subsurface horizons with very little organic matter.
They have high bulk densities, a hard or very hard
consistence, are brittle and usually have a massive or
platy structure. Horizons which are simply compacted
appear to correspond most closely to the fragipans
described by American authors and are referred to as
B3, Bx, Cx or Cgx horizons in the various horizon
nomenclatures. Continuously cemented material, described
as indurated, is cemented with one or more of the
following:- iron, aluminium, organic matter, calcium
carbonate or silica; such horizons are designated by
the letter m, e.g. Bhm, Ckm, Bmk or Bms.
     Although these compact horizons occur in many parent
materials, they are typically seen in re-sorted glacial
till. A typical field description would be:-

80-130cm   Yellowish brown (10YR5/4) prominently
                            mottled with reddish brown (5YR4/4)
                            and light brownish grey (10YR6/2);
                            compact extremely stony apedal sandy
                            clay loam; rare pores and fissures;
                            no visible organic matter; no roots;
                            iron and manganese concretions.

     The field description is supported by micromorpho-
logical description:- angular fine sand and silt grains
in a dense brown plasma with undulic/isotic fabric in
glaebules; rare voids; many lithorelicts, some with
evidence of silt cappings. SEM examination indicates
coating of amorphous silica on grains and forming simple
and compound bridges between grains.
     In Scotland different theories have been advanced
for the origin of these compact horizons. FitzPatrick
(1956) was able to show that the main features were of
periglacial origin, and Romans (1962) showed that some
indurated horizons resulted from cementation of the
lower soil horizon with alumina. Discussions which have
not distinguished between layers due simply to compaction

from those cemented by some illuvial agent, have tended
to confuse the issue. In America, the subject of fragipans
was discussed by Grossman and Carlisle (1969) who describe
fragipans as dense, brittle and rigid subsoil layers,
probably bonded by clays. Earlier, Grossman and Cline
(1957) had found that the rigidity of fragic layers
correlated highly with clay percentage, specifically
illite. The theory of clay bonding has also received
support from Wang et al., (1974) who proposed that clay
bridges between sand and silt particles gave the
characteristic brittleness.

There appears to be agreement that at least two
distinct features are present resulting from different
modes of genesis. The compaction results from former
periglacial processes, but in many cases there is a
subsequent introduction of a cementing agent. In America
the term fragipan is used to describe relict periglacial
features as well as those compact layers which do not have
silt cappings, slake on wetting and which occur well to
the south of areas known to have experienced periglacial
climates.

There are many obscure areas in our knowledge of
these compact subsurface layers as the previous discussion
indicates. With these controversies in mind, the character
of soils in south Wales which exhibit these features has
been investigated. Methods of approach have included
physical, chemical and microscopical techniques.

The physical appearance of horizons with high bulk
density in soils of south Wales is rather variable.
Colours are inherited from the parent material, usually
a glacial deposit, subjected to periglacial modification.
Structure is massive, but some platiness can be observed
also. The characteristic brittle fracture results in
irregular fragments when disturbed. These horizons are
commonly very stony, which adds to their impenetrability.
Silt cappings occur on the upper surface of the stones and
it has been implied by some investigators,e.g. FitzPatrick
(loc. cit.), Romans (loc. cit.), that the lower limit
of the compact layer is indicated by the depth to which
these features persist. Frequently frost cracks can
be observed associated with the compact layer and these
have been noted in central and south Wales. The upper
boundary of the compact layer does not appear to be as

abrupt as those in corresponding layers in Scotland
described by Romans (1962, 1976).

     Particle-size determinations indicate that the
texture of the compact material is similar to that of
horizons above, ranging from sandy loam to a sandy clay
loam, an observation which agrees with that of Rudeforth
(1976) for central Wales. There is a slight tendency for
an increased amount of silt to be present in the compact
layers and Romans (1976) provides evidence that this
enrichment by silt took place by the time Pollen Zone
3 accumulated in Scotland. Bulk density measurements
agree with those from similar soils elsewhere,being
in the range 1.90 to 2.00.

     Conventional chemical analyses of the compact
horizons in this and other studies have been inconclusive.
The present pH of the soil profiles and the compact
material lies in the range 4.0 to 5.6 Compactness caused
by freeze-thaw processes could well be contemporaneous
with the Devensian origin of the materials, but cementation by other elements must have occurred subsequently as
Catt (1979) suggests. Iron and aluminium contents in the
profiles examined are at a maximum in the friable horizons
with less in the layers of high bulk density. The total
amounts are also small and unlikely to be sufficient to
affect the material as strongly as has been claimed by
some authors. It is not thought that the cementation
is associated with alumina released with podzolisation
in post-Neolithic times as Romans (1976) claims, as
these compact and indurated horizons occur in a wide range
of different soil types.

     Examination of these compact horizons by optical
microscopy, combined with point counting, revealed a
sharp decline in voids and only weak preferred orientation
of clay minerals. In the compact material, roots were
virtually absent and there were increased numbers of
rock fragments and some unweathered grains of feldspar.
As these feldspars are absent from the soil horizons
above it suggests the compact horizon has resisted the
contemporary weathering processes, enabling the feldspars
to persist. The fabric has been described as masepic
or vo-masepic. There is little evidence of the introduction of illuvial clay material, nor of its role in

cementing material together.

Examination by SEM and analysis by the LINK system
enables high resolution to be combined with pin-point
accuracy of the electron microprobe. This combination
of microscopy and chemical analysis has much to offer
the soil micromorphologist. Several authors have used
SEM for the study of microfabrics (Eswaran, 1971; Stoops,
1974; Smart, 1974; Tovey, 1974), but none has satis-
factorily identified the cause of induration in these
horizons, and considerable controversy remains (Lynn
and Grossman, 1970; Wang et al., 1974).

 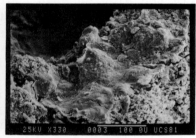

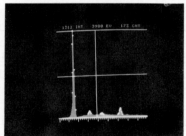

Fig. 1. (a) Fine growth of silica crystals coating
        larger skeleton grains in a matrix which has
        compound silica bridges between larger grains;
        (b) Ubiquitous crypto-crystalline silica
        coating matrix and skeleton grains; (c) X-ray
        analysis of intergranular bridge showing strong
        Si peak with weak responses for K and Fe. The
        absence of Al helps to confirm Si as an
        indurating material rather than clay or Al alone.
        Other samples examined have shown a weak
        response for Al just to the left of the Si peak
        (based on count of 5000).

Samples from these compact horizons of the soils of
south Wales possess re-precipitated crypto-crystalline
silica. Normally, this precipitation can be seen to
take the form of intergranular bridges between skeleton
grains (Fig. 1a), but in well developed examples silica
overlies both skeleton grains and matrix (Fig. 1b).
X-ray analysis of the intergranular bridges by the LINK
system gives a weak response for iron and aluminium
(if any) but a pronounced peak for silica (Fig. 1c). The
soils from which these samples were obtained include
brown soils, brown podzolic soils, cambic stagnogley
soils and gley podzols. The authors support the concept
of periglacial origin for the compact material and point
to strong regional evidence for silica as the element
responsible for the subsequent induration of the material.
   The origin of the silica is not directly associated
with podzolisation but can be ascribed to processes
operating upon the weathering drift material after its
re-deposition by solifluction. At first, with a probable
abrasion pH of around neutrality, pH values would
gradually drop as acid hydrolysis affected the mineral
material in the upper part of the profile. Also the till
would contain much finely ground material (rock flour)
which could be acted upon readily by hydrolysis or acid
hydrolysis, and in the breakdown silica would be released
(Raiswell et al., 1980). Experiments by Acquaye and
Tinsley (1965) upon silica present in soils showed
marked changes in its solubility in the range pH 4 to 6.
In the presence of ferric iron, a sharp change in silica
solubility occurred between pH4 and 5 but in the presence
of aluminium this change took place between 5 and 6. The
presence of humus also increases the solubility of
silica through the pH range 5.5 to 9.0. Thus, although
the solubility of silica is greatly reduced above pH4,
a mechanism does exist whereby it can still pass into
solution, although Paton (1978) suggests it is small in
amount and in the form of silicic acid (H_4SiO_4).
   Silica reprecipitation is known in several soil
groups including podzols, solodic soils and in silcretes.
However, little attention has been given to the transport
of silica in other soils. Acquaye and Tinsley (1965)
claim that a change of less than 0.5 of a pH unit is

sufficient to cause the reprecipitation of silica in
a podzol and there is no reason to think such differences
did not exist when these soils were undergoing their
initial stages of formation. Redeposition of silica
took place in the form of extensions to crystals, as
intergranular bridges, and in extreme cases as ubiquitous
crypto-crystalline reprecipitation over all the original
particles. Harlan et al. (1977) lend weight to this
argument in their discussion of fragipan formation in
soils of Indiana but they rest their case on bulk
analyses and not on analysis of the linking structures.
       An examination of profiles from south Wales, using
the complementary techniques of field, chemical and
microscopical investigation points to the role of
amorphous silica as the cement which has indurated
these compact horizons. In the experience of the authors,
neither clay, iron, aluminium nor organic matter are
critical although all may be present.

ACKNOWLEDGEMENTS
       The authors wish to thank Mr. Malcolm Williams
for assistance with the scanning electron microscope
and Messrs. Edward Price and Alan Cutliffe for photographic support.

REFERENCES
Acquaye, D.K. and Tinsley, J. 1965. Soluble silica in
       soils. In: E.G. Hallsworth and D.V. Crawford
       (Ed), Experimental Pedology. Butterworth, 126-148.
Bridges, E.M. and Clayden, B. 1971. Pedology. In:
       W.G.V. Balchin (Ed), Swansea and Its Region.
       British Assn. for Advancement of Science, Swansea,
       73-84.
Bull, P.A. and Bridges, E.M. 1978. Micromorphological
       and genetic properties of a gleyic brown podzolic
       soil from south Wales, U.K.   Unpub. Paper, 11th
       I.S.S.S. Conference, Edmonton, Canada.
Carlisle, F.J., Knox, E.G. and Grossman, R.B. 1957.
       Fragipan horizons in New York Soils: General
       character and distribution. Soil Sci. Soc. Amer.
       Proc., 21, 320-321.

Catt, J.A. 1979. Soils and Quaternary geology in
     Britain. J. Soil Sci., 30, 607-642.
Clayden, B. and Evans, G.D. 1974. Soils in Dyfed I:
     Sheet SN41 (Llangendeirne). Soil Survey Record No.
     20. Soil Survey of England and Wales, Harpenden,
     England.
Crampton, C.B. 1965. An indurated horizon in soils of
     south Wales. J. Soil Sci., 16, 230-241.
Eswaran, H. 1971. Electron scanning studies of the
     fabric of fracture surfaces. Soil Sci. Soc. Amer.
     Proc., 35, 787-790.
FitzPatrick, E.A. 1956. An indurated soil horizon
     formed by permafrost. J. Soil Sci., 7, 248-254.
FitzPatrick, E.A. 1971. Pedology. Oliver and Boyd,
     Edinburgh.
FitzPatrick, E.A. 1976. Cryons and Isons. Proc. of
     North of England Soil Disc. Gp. Report No.11,
     31-43.
FitzPatrick, E.A. 1980. Soils. Longman.
Grossman, R.B. and Cline, M.G. 1957. Fragipan horizons
     in New York Soils 2: Relationships between rigidity
     and particle size distribution. Soil Sci. Soc. Amer.
     Proc., 21, 322-325.
Grossman, R.B. and Carlisle, F.J. 1969. Fragipan soils of
     the Eastern United States. Adv. Agron. 21, 237-279.
Harlan, P.W., Franzmeier, D.P. and Roth, C.B. 1977. Soil
     formation in loess in southwestern Indiana. II.
     Distribution of clay and free oxides in fragipan
     formation. Soil Sci. Soc. Amer. Proc., 41, 99-103.
Hodgson, J.M. (Ed), 1974. Soil Survey Field Handbook.
     Tech. Monogr. No.5. Soil Survey of England and
     Wales, Harpenden, England.
Lynn, W.C. and Grossman, R.B. 1970. Observations of
     certain soil fabrics with the scanning electron
     microscope. Soil Sci. Soc. Amer. Proc., 34, 645-648.
Matthews, B. 1976. Soils with discontinuous induration
     in the Penrith area of Cumbria. Proc. North of
     England Soil Disc. Gp. Report No.11, 11-19.
Paton, T.R. 1978. The Formation of Soil Material.
     George Allen and Unwin.
Raiswell, R.W., Brimblecombe, P., Dent, D.L. and Liss,
     P.S. 1980. Environmental Chemistry. Edward Arnold.

Romans, J.C.C. 1976.   Indurated layers.   Proc. North
     of England Soil Disc. Gp. Report No.11, 20-30.
Romans, J.C.C. 1962.   The origin of the indurated B3
     horizon of podzolic soils in north-east Scotland.
     J. Soil Sci., 13, 141-147.
Rudeforth, C.C. 1976.   Soil features associated with
     former periglacial conditions in west and central
     Wales.   Proc. North of England Soil Disc. Gp.
     Report No.11, 44-50.
Smart, P. 1974.   Electron microscope methods in soil
     micromorphology.   In: G.K. Rutherford (Ed),
     Soil Microscopy.   The Limestone Press, Kingston,
     Ontario, 190-206.
Soil Survey Staff. 1975.   Soil Taxonomy.   Agricultural
     Handbook 436, U.S.D.A. Washington, D.C.
Stewart, V.I. 1961.   A permafrost horizon in the soils
     of Cardiganshire.   Welsh Soils Disc. Gp. Report No.
     2, 19-22.
Stoops, G. 1974.   Optical and electron microscopy.   A
     comparison of their principles and their use in
     micropedology.   In: G.K. Rutherford (Ed),   Soil
     Microscopy.   The Limestone Press, Kingston, Ontario.
     101-118.
Tovey, N.K. 1974.   Some applications of electron
     microscopy to soil engineering.   In: G.K. Rutherford
     (Ed),   Soil Microscopy.   The Limestone Press,
     Kingston, Ontario.   119-142.
Wang, C., Nowland, J.L. and Kodama, H. 1974.   Properties
     of two fragipan soils in Nova Scotia including
     scanning electron micrographs.   Can. J. Soil Sci.,
     54, 159-170.
Wright, P.S. 1980.   Soils in Dyfed IV: Sheet SN62
     (Llandeilo).   Soil Survey Record No.61.   Soil
     Survey of England and Wales, Harpenden, England.

A MICROSCOPIC STUDY OF QUARTZ EVOLUTION IN ARID AREAS

A. Halitim, M. Robert and J. Berrier

I.N.R.A., Station de Science du Sol, 78000 Versailles,
France

ABSTRACT
     Using optical and scanning electron microscopy,
the morphology and nature of the surface of quartz
grains in various environments and the relationships
between quartz, calcite and palygorskite have been
examined. A dynamic system for silica in arid environments appears to exist. When quartz undergoes invasion
by calcite, epigeny patterns may be produced. Neoformation of palygorskite is related to the dynamic
system of silica.

INTRODUCTION
     Two environmental factors in arid areas are
particularly important: (1) continuous and intense
wind action producing considerable aeolian accumulations; (2) the effect of salts. Carbonates also seem
to play a significant role in so far as they are involved
in quartz epigenesis (Nahon et al., 1975; Delhoume,
1980) and in palygorskite neogenesis (Millot et al.,
1977).
     Using optical and scanning electron microscopy
(SEM), the morphology of quartz grains in various
environments was examined as well as the relationships
between quartz, calcite and palygorskite.

MATERIAL AND METHODS
     The material was collected in the central part
of the high steppes in Algeria (Fig. 1a). This area
is very windy and has an average annual rainfall between
200 and 250 mm.
1. Quartz grains were separated from samples from six
pedological situations including calcareous crusts and
nodules and non-calcareous soils developed in Mio-
Pliocene red clay. The grains, which were separated
without the removal of carbonates, were examined with

A. Halitim et al.

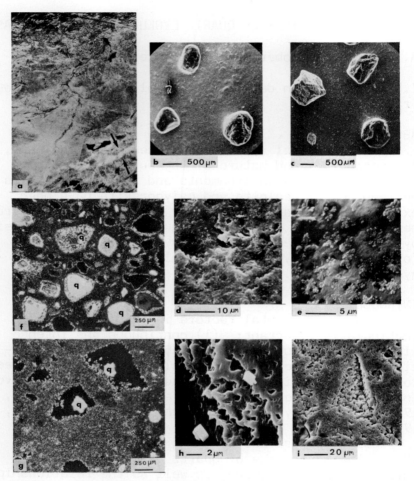

Fig. 1. (a) Observation area of the Zahrez dune belt:
the arrows point to the dune belt (1:1,000,000).
Aeolian quartz: (b) Round mat grains; (d) Characteristic surface appearance, powdery deposit
with splinters and scales; (e) Flower-like
silica deposits.
Quartz in calcareous accumulations: (c) Rounded
edged grains; (f) Various patterns of calcite
penetration into quartz; (g) Quartz in voids;
(h) Very powdery surface with calcite crystals;
(i) Powdery deposits in impact furrows.

a binocular microscope. Some of the quartz minerals
were ultrasonically washed to clean their surfaces
prior to examination. In addition, soil and rock
fragments were examined for quartz-calcite-clay relationships. After fixation on stubs and coating with
Au-Pd,all the specimens were observed using a SEM
Jeol JSM 35.
2. Six sand samples, collected at various depths in the
dune belt, were ultrasonically washed and observed with
SEM.

RESULTS AND DISCUSSION
Quartz Features in the Dune Belt
     Among aeolian formations in this area, the Zahrez
dune belt (Fig. 1a) is the most important of the
Quaternary formation due to its size. According to Pouget
(1969), sandstone outcrops in the Saharian Atlas are the
main source of the sand.
     The quartz minerals of the dune belt have a highly
characteristic morphology. They are well rounded (Fig.
1b) (Krinsley and McCoy, 1978) and their surfaces are
covered with scales and crescent-like impact patterns.
Surface cracks with silica 'flower-like' precipitation
can be observed on some grains (Fig. 1e).
     The shape of grains and the crescent-like de-
pressions and scales occur as a result of saltation,
wind transport and collisions between grains. Cracks
and silica precipitation are attributable to chemical
processes. It is likely that such precipitation
occurred once sand had reached the dune belt, as a
relatively stable condition is required to crystallise
silica into 'flower-like forms' and maintain this
crystallisation.
     Silica dissolution and reprecipitation might result
from dew phenomena. In these arid areas, the relatively
important dew can penetrate 2-3 cm into the dune (Engel
and Sharp, 1958). Water containing CO_2 evaporates by
day and the water film pH increases due the presence
of dissolved evaporites (Krinsley and Doornkamp, 1973),
thus leading to solution of silica. The solubility of
silica is actually increased due to the presence of a
powdery layer resulting from abrasion during the migration
period.

A. Halitim et al.

    The silica usually precipitates when mixed with
other cations and salts (KCl, Al^{+++}, Fe^{+++}, Mg^{++}), which
can be identified with a microprobe. According to
Le Ribault (1977) the presence of 'silica flower-like
forms' is due to a minor discontinuous deposit.
    In summary, mechanical phenomena prevail in the
belt, and result in microscopic effects of silica
dissolution and reprecipitation.

Quartz Evolution in Calcareous Accumulations
    (a) Quartz-calcite relationships:
Using the optical microscope quartz grains of variable
size surrounded by a ring of calcite microcrystals can
be observed. Some of these quartz minerals show etch
pits and may be spotted with calcite microcrystals
(Fig. 1f). Calcite concentrations in voids appear to
have replaced previous quartz crystals. Finally, in
some areas with fewer carbonates, quartz crystals are
surrounded to a large extent by a void (Fig. 1g).
    SEM shows that quartz minerals in the calcareous
formations are totally different from those in the dune
belt, as chemical processes prevail. These chemical
actions can occur on quartz that has previously been
subjected to aeolian effects (impact crescent and round
shape) or to Mio-Pliocene fluvial action (rounded edges,
Fig. 1c and impact furrows).
    The surface of quartz crystals is often powdery
(Fig. 1h) especially in depressions from previous impact
furrows (Figs. 1i and 2a).
    In some cases, it can be seen that quartz is partly
converted into small crystals, and that needle-shaped
calcite is completely mixed with the quartz (Fig. 2b)
and seems to grow inside the powdery zone. These
needles can concentrate in cavities.
    It seems also that some dissolution patterns
correspond to calcite rhombohedra. In some cases,
calcite substitution for quartz is such that the two
can hardly be distinguished (Fig. 2c and d).
    Silica dynamics in such an environment can be
illustrated by flower-like forms precipitated in some
quartz minerals (Fig. 2e). Nevertheless these flower-
like forms were found only on quartz minerals in voids.

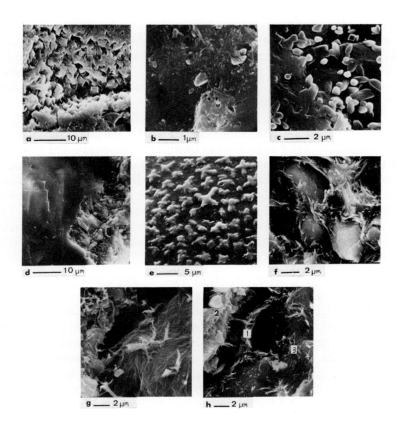

Fig. 2. (a) Higher magnification of powdery deposits
shown in Fig. 1 i ; (b) Needle calcite on
powdery quartz surface; (c) Calcite crystals
on the quartz surface; (d) Calcite substitution
for quartz; (e) Silica 'flower-like' deposits
rich in KCl; (f) Calcareous accumulations
very rich in palygorskite which envelops
calcite crystals; (g) Calcareous crust poor in
palygorskite, quartz grain enclosed by palygorskite; (h) Calcareous crust poor in palygorskite,
clay (1) separates quartz (2) from calcite (3).

A. Halitim et al.

Powder patterns and flower-like deposits are likely
to reflect some previous silica solution. Obviously,
silica release in these areas can only be explained by a
high pH due to the presence of calcium carbonate and,
probably, of Mg^{++}. Chemical analysis shows that such
carbonate-rich media may also be rich in Mg^{++}.

(b) Quartz-calcite-palygorskite relationships:
Palygorskite is fairly frequently found in arid and semi-
arid areas, especially in calcareous accumulations.
However, various studies of the origin of this mineral
have not yet explained its neogenesis. Some authors, like
Wiersma (1970), believe that palygorskite results from
slight chemical alteration of minerals. Others, however,
think that it is neoformed in calcareous crusts (Millot
et al., 1977).

SEM is a useful method for observing palygorskite
neogenesis in particular sites. When observing
fragments of soils and rocks rich in palygorskite, it
can be seen that this clay encloses calcite grains as
well as quartz crystals (Fig. 2f). It seems, though,
that it was formed in situ after the sediments had been
deposited, if fibre length, fibre packing and the
absence of short and broken fibres are considered. It
should be noted that the amounts of silica and magnesium
required for such neoformation can be present in the
environment (15 to 25 ppm of SiO_2, 100 to 200 ppm of
Mg^{++} in 1/1 soil extracts).
By contrast, in calcareous environments poor in
palygorskite (where palygorskite cannot be identified
by X-rays), this materal encloses only a few quartz
grains (Fig. 2g).   Palygorskite even separates dissolving
quartz from calcite (Fig. 2h). In this case it seems
logical to interpret the observations as follows :
carbonates allow release of silica, which precipitates
with Mg^{++} and possibly also with Fe^{+++}, when present,
to give a fibrous clay.

CONCLUSIONS
This morphological study shows the existence of
silica dissolution and reprecipitation in arid areas.
This process is a feature of the dune belt sand as a

result of dew and salt action.

In calcareous accumulations, silica dissolution
and reprecipitation could be observed as well as calcite
substitution for quartz, producing in some cases
epigenetic patterns. Such carbonate action was
experimentally verified by the senior author who found
that a $CaCO_3$ saturated solution dissolved 5 to 6 times
more silica from quartz than water and neutral salts.

Finally, it is likely that the silica released
partly promotes palygorskite neoformation when other
conditions are met, such as sufficient Mg^{++} (and perhaps
Fe^{+++}) and high pH, which is often the case in such
areas.

REFERENCES
Delhoume, J.P. 1980. L'épigénie calcaire en milieu
     méditerranéen semi-aride. Colloque sur les
     carbonates, Bordeaux, 163-171.
Engel, C.G. and Sharp, R.P. 1958. Chemical data on
     desert varnish. Bull. Geol. Soc. Am.,69, 487-518.
Krinsley, D.H. and Doornkamp, J.C. 1973. Atlas of
     quartz sand surface textures. Cambridge University
     Press, 91 pp.
Krinsley, D.H. and McCoy, F. 1978. Aeolian quartz sand
     and silt. In: SEM in the study of sediments. Geo.
     Abstracts Ltd., 249-260.
Le Ribault, L. 1977. L'Exoscopie des Quartz. Masson,
     150 pp.
Millot, G., Nahon, D., Paquet, H., Ruellan, A. and Tardy,
     Y. 1977. L'épigénie calcaire des roches silicatées
     dans les encroûtements carbonatés en pays subaride
     Antiatlas Maroc. Sci. Géol. Bull., 30, 129-152.
Nahon, D., Paquet, H., Ruellan, A. and Millot, G. 1975.
     Encroûtements calcaires dans les altérations des
     marnes éocènes de la falaise de Thiès (Sénégal).
     Organisation morphologique et minéralogique. Sci.
     Géol. Bull., 28, 29-46.
Pouget, M. 1979. Les relations sol-végétation dans les
     steppes sud-algéroises (Algérie). Thèse, Marseille,
     466 pp.
Wiersma, J. 1970. Provenance, genesis and paleogeo-
     graphical implications of microminerals occurring
     in sedimentary rocks of the Jordan valley area.
     Thèse, Ph.D. Amsterdam.

CALCAREOUS CRUST (CALICHE) GENESIS IN SOME MOLLISOLS AND ALFISOLS FROM SOUTHERN ITALY : A MICROMORPHOLOGICAL APPROACH

D. Magaldi

Istituto Sperimentale per lo Studio e la Difesa del Suolo, Firenze, Italy

ABSTRACT
     Micromorphological studies carried out on calcareous crusts (caliche) from some Mollisols and Alfisols observed respectively in Puglia and Sardinia, South Italy, show that the pedogenic process of calcareous crust formation differed between the two groups of soils. In the Mollisols the process involving calcium super-saturated solutions occurred near the soil surface in medium textured horizons and produced a micritic crust with banded structure and high porosity; this process is still active today. In Alfisols calcium carbonate deposition originated in deep coarse-textured horizons from initially sub-saturated waters which formed micro-sparry and undifferentiated crusts with very low porosity in superimposed layers; this process is not active at present.

INTRODUCTION
     Some soils from the Italian Mediterranean region have horizons with calcareous accumulations (caliche), at a depth varying from about 50 cm to 2 m, which at times are very hard, laminated or nodular (Mancini et al., 1966).
     To determine the origin of the crust and its different aspects, two sets of calcareous crusts from 10 profiles of Mollisols and 3 profiles of Alfisols were studied in the field and in the laboratory. Ten samples from each set were selected for analysis.

MATERIALS AND METHODS
     Samples of Mollisol crusts were taken from Tavoliere, a large plain of marine regression in northern Puglia (mean latitude and longitude of sampling area: N 41°30' and E 15°30').

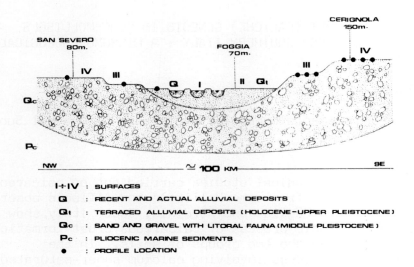

I + IV : SURFACES
Q   : RECENT AND ACTUAL ALLUVIAL DEPOSITS
Qt   : TERRACED ALLUVIAL DEPOSITS (HOLOCENE-UPPER PLEISTOCENE)
Qc   : SAND AND GRAVEL WITH LITORAL FAUNA (MIDDLE PLEISTOCENE)
Pc   : PLIOCENIC MARINE SEDIMENTS
•   : PROFILE LOCATION

Fig. 1. Geomorphological setting of the crust samples
        from Mollisols.

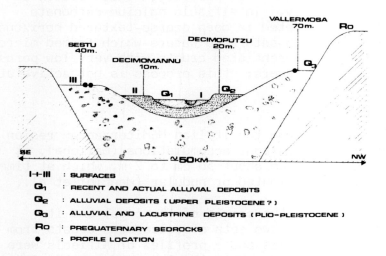

I + III : SURFACES
Q1   : RECENT AND ACTUAL ALLUVIAL DEPOSITS
Q2   : ALLUVIAL DEPOSITS (UPPER PLEISTOCENE?)
Q3   : ALLUVIAL AND LACUSTRINE DEPOSITS (PLIO-PLEISTOCENE)
Ro   : PREQUATERNARY BEDROCKS
•   : PROFILE LOCATION

Fig. 2. Geomorphological setting of the crust samples
        from Alfisols.

The climate is xero-thermomediterranean (according to Emberger, 1955) with average precipitation of 460 mm and mean annual temperature around 16°C.

According to Soil Taxonomy (Soil Survey Staff, 1975), the soils are classified as Petrocalcic Palexerolls (9 profiles) and Typic Calcixerolls (1 profile). The former are found on marine calcareous sand with some inter-calation of calcareous and siliceous gravel from the early Quaternary and the latter on more recent gravelly alluvial deposits (Fig. 1). All the soils are ploughed to the top of the crust, located at about 50 cm below the surface. The typical horizon sequence is Ap/Cca or Ap/Ccam.

The Ap horizon may be more or less calcified, pH varying from 7.5-8.5. The $CaCO_3$ content reaches a maximum of 30%. These soils are moderately fine textured.

The crust samples collected from Alfisols come from Campidano, a vast alluvial plain in Southern Sardinia (mean lat. and long. of sampling area: N 39°20' and E 9°00').

The climate is thermomediterranean with average annual precipitation around 550mm and mean annual tempera-ture of 15°C.

The soils sampled are coarse- and medium-textured and decalcified (pH 6.0-6.5) and can be classified as Calcic and Petrocalcic Palexeralfs. They have calcareous accumulations at a depth varying from 1-2 m, often in superimposed layers. The 3 profiles are found on a terraced plain, slightly inclined SE, formed predominantly of siliceous gravelly and sandy alluvial sediments from the Lower Pleistocene (Fig. 2).

Some macromorphological features of the crusts and the parent materials are shown in Table 1.

The micromorphological observations and apparent porosity assessment were carried out on sections of crusts with a polarising microscope and Leitz Classimat image analyser respectively.

The granulometric analyses were made by Coulter Counter after disaggregation of the sample with prolonged ultrasonic treatment. For each granulometric distribution the mean size was calculated according to the formula $Mz = \left[\emptyset_{84} + \emptyset_{16} + \emptyset_{50}\right]/3$ (Folk and Ward, 1957), and the

Table 1. Macroscopic aspects of calcareous crust samples.

| | CALCAREOUS CRUST (CALICHE) | | PARENT MATERIAL | |
	Type	Colour (hue)	Texture	Lithological type
MOLLISOLS	Mainly laminated crusts ; seldom nodular crusts	Mainly 10 YR ; seldom 7.5 YR	Sand and sand plus gravel in the same proportions	Mainly calcareous
ALFISOLS	Mainly massive crusts ; less often nodular crusts	Mainly 7.5 YR ; seldom 5 YR and 10 YR	Mainly sand plus gravel	Mainly non-calcareous

sorting coefficient according to the formula
So $= [\emptyset_{84} - \emptyset_{16}]/2$ (Inman, 1952).

TERMINOLOGY

Micromorphological descriptions were made using mainly the terminology of Brewer (1964) with some modifications based on results of more recent pedological and sedimentological studies (Choquette and Prayl, 1970; James, 1972; Bal, 1975a and b; Klappa, 1979). Revised terms used and their definitions are as follows:-

a) Ca-matrix (= calcitic matrix): The calcitic material is composed of more or less similar crystalline or cryptocrystalline particles and includes mono- or poly-units distinguishable from surrounding material by shape, composition, boundary and size. Ca-matrix types are: (1) micritic (particles $<10\mu m$); (2) microsparry (particles $10-50\mu m$); (3) sparry (particles $>50\mu m$). Ca-matrix arrangements are: random (the particles are distributed randomly) and nodular (the particles are concentrated in clusters which display a nodular shape).

b) Ca-features: Morphological units with mainly calcareous composition occurring in Ca-matrix. The Ca-features are: (1) Ca-nodules (calcitic nodules of micritic type which can be orthic, disorthic and allothic (Wieder and Yaalon, 1974), (2) Ca-concretions (calcitic concretions, after Brewer, 1964) and (3) Ca-crystallaria (calcitic crystallaria after Brewer 1964). Ca-crystallaria have microsparry or sparry crystals and can be subdivided into chambers, tubes, intercalary crystals and needle crystals (acicular crystals, lublinite). The needle crystals are considered a distinguishing feature of the crystallaria because of their significance in micro-environment definition. Intercalary crystals are different because of their habits which can be equant or prolate.

c) Voids (Fig.3): A new morphological voids classification is used: (1) irregular voids (mainly vughs according to Brewer's concept and others with irregular shape), (2) single fractures (joint and skew planes of Brewer's classification), (3) composite fractures (mainly craze planes, after Brewer, and other anastomosing voids) and (4) skeletal voids (equant voids formed by decayed plant

D. Magaldi

Table 2.   Micromorphological and other analyses.

	MOLLISOLS		ALFISOLS	
	Micromorphological analyses			
Skeleton grains	S > L = 7 % (8 samples) S = L = 1 % (1 sample) S < L = 5 % (1 sample)		S > L = 8 % (3 samples) S < L = 1 % (7 samples)	
Ca-matrix type	micrite (in all samples)		micrite (5 samples) microspar (5 samples)	
Ca-matrix basic distribution	nodular > random		random > nodular	
Voids	many irregular voids ; common skeletal voids and composite fractures		many simple fractures ; common irregular voids and skeletal voids	
Ca-nodules	many nodules ; orthic > disorthic		common nodules; orthic nodules only	
Ca-crystallaria	common crystallaria ; many needle crystals; crystal tubes and chambers > intercalary crystal		scarce crystallaria ; needle crystals absent ; intercalary crystals > crystal tubes and chambers	
Cutans	scarce calcitans ; absent or very rare argillans		common calcitans and argillans	
Ca-structure	banded > undifferentiated scarce brecciated		undifferentiated > banded very rare brecciated	
	Other analyses			
	$\bar{x}$	s	$\bar{x}$	s
CaCO$_3$ %	65	9	65	14
Apparent porosity % (in thin section)	22	19	10	10
Mean diameter (Mz) (phi units)	6.8	0.3	6.6	0.5
Sorting (So) (phi units)	1.8	0.4	1.6	0.3

roots and/or vanished skeleton grains).
d) Ca-structure (= calcareous crust structure):
   Corresponds to primary structure after Brewer's classification and can be subdivided into three broad groups:
   (1) undifferentiated (homogeneous crust as to size,
   shape and arrangement of pedological and Ca-features),
   (2) banded (laminated crust with different crystal size
   in individual laminae) and (3) brecciated (fragmented
   crust into an irregular network of fractures and clasts).

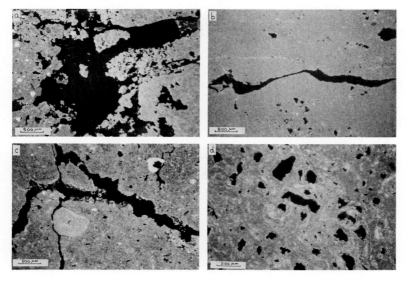

Fig. 3. Morphological classification of voids used in
   this paper: (a) Irregular voids; (b) Simple
   fractures; (c) Composite fractures; (d) Skeletal
   voids.

RESULTS
   The results of the analyses are reported in Table 2.
   Calcimetry: The CaCO₃ content is essentially the
   same in all the samples. Slight variations are due
   to different quantities of skeleton grains embedded
   in the crust.
   Porosity: This is noticeably higher in Mollisols than
   in the Alfisols. The morphology of the voids is also
   different in the two groups.

Granulometry: The average values are approximately
the same in the two sets of samples. This shows
that the growth of the crystals is on average the
same in the two groups. The distribution diagram of
the average values corresponding to all the samples is
bimodal with a separation of around 13 μm between the
two maxima (Fig. 4). The samples with Mz lower than
13 μm correspond to micrites while those with greater
Mz are microspars. This result emphasises the suita-
bility of choosing 10 μm as the boundary between
microspars and micrites.
Micromorphology: In addition to the semi-quantitative
data in Table 2, the following observations were made
for each of the two groups of soils.

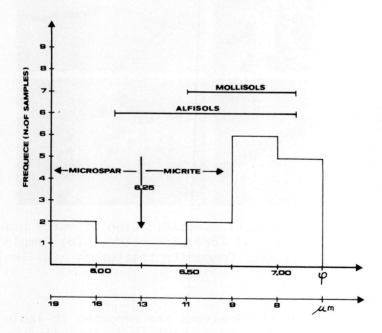

Fig. 4.   Frequency distribution of mean size of
          calcitic crystals in all crust samples (Mollisols
          and Alfisols). Note that 68% of the observed
          values are distributed in the range indicated by
          the horizontal bars.

a) Mollisols: The Ca-matrix (see earlier definition)
is always micritic although with some tendency towards
microsparitic. The matrix is usually dark and is sometimes
opaque between crossed polarisers. The basic distribution
of the calcite grains is random to nodular. Most of the
silicate skeleton grains (quartz, feldspars, augite,etc.)
are dispersed in the calcareous mass (floating grains) and
show signs of dissolution at the edges (corroded grains);
they are sometimes fragmented. The Ca-nodules sometimes
have a boundary consisting of concentric zones of dark
micrite, sometimes a diffuse boundary, without particular
arrangement of the constituents. The crystallaria consist
of subrounded calcite crystals with rhombohedral habit
together with more elongated crystals. In most samples
needle-shaped calcite crystals occur, similar in shape and
aspect to those already described in the literature
(Parfenova and Yarilova, 1965; Sehgal and Stoops, 1972;
James, 1972; Bal, 1975a; Klappa, 1979; Knox, 1977).
     These crystals are woven (woven fabric, hyphantic
fabric according to Bal, 1975a;   acicular networks
according to Knox, 1977) and are often found in the pores
that pass through the crust. Crystal chambers more or less
infilled with sparry calcite crystals,sometimes with
geopetal arrangement,are also common.
     The most frequent structure is a banded one. The bands
are formed by micrite with a nodular arrangement alternating
with microsparry laminae, at times having undulations of an
anticlinal and synclinal type. Crusts without banding are
less frequent.
     In places fragmentation of the crust has taken place
with subsequent welding and embedding of the fragments
by sparry calcite (brecciated structure).
b) Alfisols: The Ca-matrix is either micritic or microsparry with rounded crystals, at times in the form of
aggregations of clearly defined sparry calcite. The
matrix colour is grey or more rarely darker. The basic
distribution of the crystals is random, tending only in
a few cases to nodular. There are few silicate skeletal
grains. The Ca-nodules are similar to those in Mollisols
but more frequently have a concentric zoning around the
edges. The crystallaria, which are generally scarce,
consist of prolate intercalary crystals, crystal tubes
filled with sparry calcite and crystal chambers.

Among the Ca-concretions, welded spherulites of two or three single crystals are present;   calcite needles are absent.   Cutans consist mostly of ferri-argillans and secondly of compound cutans (a calcitan layer and a ferri-argillan layer).   Calcitans are composed of elongate calcitic crystals arranged with the large axis perpendicular to the void and are nearly always covered by the argillans which are sometimes finely stratified and sometimes strongly degraded.   Many argillans are embedded in the Ca-matrix in the form of papules.

The structure of the crusts is predominantly un-differentiated;   rarely it is banded and very rarely brecciated.

CONCLUSIONS

There are some morphological differences between the calcareous crusts observed in the Mollisols and in the Alfisols, corresponding to different pedogenetic processes.   The following models are proposed :

In the Mollisol crust, formation appears to take place because water rich in Ca and Mg circulates slowly allowing crystallisation to occur.   According to James (1972) and Folk (1974), calcite precipitation from waters supersaturated with Ca and Mg inhibits the growth of large calcite crystals and leads to the formation of needle-shaped crystals with a more or less woven arrangement. This woven arrangement constitutes the first stage in crust genesis.   Subsequently, by means of a dissolution/ recrystallisation process (James, 1972;   Sehgal and Stoops, 1972;   Esteban, 1974) more or less distinct micritic calcite nodules appear.   Crystal growth, and hence the tendency towards microspar, is slowed down by a fine texture and by a strong degree of saturation of the circulating solutions.   Organic matter and silicate grains are embedded in the Ca-matrix during the crystal growth (James, 1972).   Because of the shallow depth at which these crusts occur, periodic soil drying is likely to contribute to hardening of carbonates and to irregular disruption of the crust (brecciated structure). Waters less rich in Ca due to progressive profile decalcification circulated.   Evaporation of the solutions gives rise to slow precipitation of sparry calcite crystals in voids and the formation of chambers, tubes

and intercalary crystals. A similar hypothesis is
advanced by Bal (1975b) for some Russian chernozems.
     Crust formation takes place either upwards or
downwards. At a later stage the waters have difficulty
in penetrating the crust and flow along its surface.
Such surface flow is aided by the particular landscape
in which these soils developed. The flow of water supersaturated in Ca near the soil surface causes rapid
deposition of wavy calcitic laminae (pellicule rubanée:
Ruellan, 1968) according to hypotheses advanced by various
authors (Ruellan, 1968; Durand, 1975; Capolini and
Sary, 1975; Braithwaite, 1979).
     Although the crust may grow upwards, some percolation
still takes place downwards because of fragmentation and
the frequency of very large voids. The presence of
needle-shaped calcite crystals in pores with remains of
roots, coupled with field observations, suggests that
crust formation is still active in the Mollisols from
Puglia and has not changed substantially since the early
Quaternary. In effect, no significant morphological
differences occur between crusts belonging to surfaces of
different age. Only those on the highest surface have
more reddening and traces of clay cutans, demonstrating
more advanced pedogenesis.
     Crusting in Alfisols takes place through water per-
colating through a coarser and therefore more porous
soil material. Callot et al. (1978) showed that in
waters often under-saturated with Ca (and Mg), the pH
may be very high (c.10) although varying widely, and
crystal precipitation therefore takes place slowly
and in the absence of ions which inhibit the normal
growth of the calcite crystals (Buckley, 1951). These
processes lead to the following: formation of larger
crystals but with some nodules; much reduction in
skeletal silica due to high pH values (according to the
relationship between pH and solubility of silica);
progressive diminution of porosity. Because of the
depth at which these solutions begin to encrust, periodic
drying causes little or no fragmentation of the crust.
A reduction in porosity is probably brought about by
subsequent illuviation of clay; plugging of the crust
causes surface flow water. Because of the slow crystallisation of the solutions, several distinct super-

D. Magaldi

imposed crusts are formed, all with similar features.
    In these soils the crusting process appears now to
have finished though limited dissolution and precipitation may still occur because of periodic drying and
wetting of the soil.

ACKNOWLEDGEMENTS
    The author wishes to express hearty thanks to
Prof. A.Aru of the Cagliari University and to Prof. P.
Baldaccini of the Sassari University for their
logistic assistance in Sardinia and for much information
on the pedological aspects of this area; the author
is also indebted to Dr. M. Pagliai of C.N.R. Laboratorio
per la Chimica del Terreno of Pisa, for his most valuable
contribution in analysing the samples with the Leitz
Classimat.

REFERENCES
Bal, L. 1975a. Carbonate in soil: A theoretical
    consideration on, and proposal for, its fabric
    analysis. 1. Crystic, calcic and fibrous. Neth.
    J. Agric. Sci., 23, 18-35.
Bal, L. 1975b. Carbonate in soil: A theoretical considera-
    tion on, and proposal for, its fabric analysis. 2.
    Crystal tubes, intercalary crystals, K fabric. Neth.
    J. Agric. Sci., 23, 163-176.
Braithwaite, C.J.R. 1979. Crystal textures of recent
    fluvial pisolites and laminated crystalline crusts in
    Dyfed, South Wales. J. Sed. Pet., 49, 190-193.
Brewer, R. 1964. Fabric and Mineral Analysis of Soils.
    J. Wiley Inc., New York. 470 pp.
Buckley, H.E. 1951. Crystal Growth. J. Wiley & Sons, Inc.
    New York, 571 pp.
Callot, G., Chamayou, H. and Dupuis, M. 1978. Variations
    du pH de la solution de matériaux calcaires en
    rélation avec la dynamique de l'eau. Eléments
    d'analyse d'un système carbonaté. Ann. Agron.,
    29, 37-57.
Capolini, J. and Sary, M. 1975. Quelques aspects de la
    répartition des croûtes et encroûtements calcaires
    et calco-gypseux du piedmont sud du Hodna. Actes du
    colloque'Types de croûtes calcaires et leur reparti-
    tion régionale'. Strasbourg, January 1975, 9-11.

Choquette, P.W. and Prayl, C. 1970. Geologic nomen-
    clature and classifications of porosity in sedi-
    mentary carbonates. Bull. Amer. Petroleum Geol.,
    54, 207-250.
Durand, I.H. 1975. Les croûtes calcaires en Algérie.
    Actes du colloque`Types des croûtes calcaires et
    leur répartition régionales'. Strasbourg, 9-11,
    January 1975.
Emberger, L. 1955. Une classification biogeographique
    des climats. Rec. Trav. Lab. Bot. Geol. Zool.
    Univ. Montpellier, 7 : 3-43.
Esteban Cerda, M. 1974. Caracteristicas de les suelos
    carbonatodos semiarides: el problema del caliche
    y sus productos asociados. Atti Seminario
    Internazionale sulla Valutazione delle Terre delle
    zone aride e semiaride dell'America Latina,
    I.I.L.A., Roma, 1974.
Folk, R.L. 1974. The natural history of crystalline
    calcium carbonate: effect of magnesium content
    and salinity. J. Sed. Pet., 44, 40-53.
Folk, R.L. and Ward, W. 1957. Brazos River bar : a
    study in the significance of grain size parameters.
    J. Sed. Pet., 27, 3-26.
Inman, D.L. 1952. Measures for describing the size
    distribution of sediments. J. Sed. Pet., 22,
    125-145.
James, N.P. 1972. Holocene and Pleistocene calcareous
    crust (caliche) profiles: criteria for subaerial
    exposure. J. Sed. Pet., 42, 817-836.
Klappa, C.F. 1979. Calcified filaments in Quaternary
    calcretes: organo-mineral interaction in the
    subaerial vadose environment. J. Sed. Pet., 49,
    955-968.
Knox, G.N. 1977. Caliche profile formation, Saldanha
    Bay (South Africa). Sedimentology, 24, 657-674.
Mancini, F. et al. 1966. Carta dei Suoli d'Italia alla
    scala 1:1,000,000, con breve commento. Comitato per
    la carta dei Suoli, Firenze.
Parfenova, E.I. and Yarilova, E.A. 1965. Mineralogical
    Investigation in Soil Science. (Translated from
    Russian). Israel Program for Scientific Transla-
    tions, Jerusalem, 178 pp.

Ruellan, A. 1968. Les horizons d'individualization et
    d'accumulation du calcaire dans les sols du Moroc.
    Trans. 9th Int. Congr. Soil Sci., Adelaide, 1968,
    Vol.4, 501-510.
Sehgal, I.L. and Stoops, G. 1972. Pedogenic calcite
    accumulation in arid and semi-arid regions of the
    indo-gangetic alluvial plain of erstwhile Punjab
    (India) - Their morphology and origin. Geoderma,
    8, 59-72.
Soil Survey Staff, 1975. Soil Taxonomy. A basic system
    of soil classification for making and interpretating
     soil surveys. Agr. Handbook No.436. Soil Conserva-
    tion Service, U.S.D.A., Washington, D.C.
Wieder, M. and Yaalon, D.H. 1974. Effect of matrix
    composition on carbonate nodule crystallization.
    Geoderma, 11, 95-121.

MICROMORPHOLOGY OF CALCRETES IN A SLOPE DEPOSIT IN THE POITEVINE PLAIN, FRANCE

J. Ducloux and P. Butel

Lab. de Pédologie, Faculté des Sciences, Poitiers, France

ABSTRACT
    Carbonate accumulations in a periglacial slope deposit
in western France are described. The pedogenic formations
are strongly differentiated with the following superimposed
horizons: 'encroûtement massif', 'pellicule rubanée',
'croûte feuilletée', 'encroûtement massif', 'accumulation
discontinué' and finally 'accumulation diffuse'.
    Accumulations of authigenic carbonates were separated
into the following types: sparitic and micritic crystals,
three sorts of needles, a micritic cement, fibrous
associations, sphaerules and veils. Observations are made
on the relationship between morphology of carbonate
concentrations, their mineralogical nature and particular
horizons.

INTRODUCTION

    Carbonate accumulations in soils particularly in arid
and semi-arid climates have been the subject of much
research (see e.g. Gile, 1961; Gile et al., 1965;
Ruellan, 1970). Such accumulations are known to occur
also in wet temperate climates, especially in western
France, in which there is a period of summer dryness
lasting for about two months.
    In the Poitevine plain, carbonate accumulations are
particularly well developed in chalky periglacial slope
deposits. This paper provides an inventory of different
types of accumulation and the mineralogical characteristics
of each in an extensive quarry section near Poitiers,
France.

THE SITE AND MATERIAL

    A gravel pit in a slope deposit at Vouillé near
Poitiers was selected for the study (Fig. 1a). The
section chosen was 2 m deep and 30 m long. It consists of
more or less regular layers of calcareous gravels, some
clay-rich, over clayey impermeable beds.

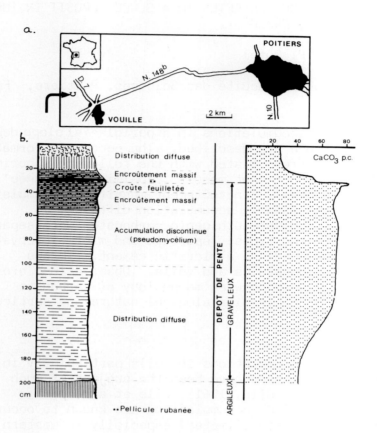

Fig. 1. (a) Location of sampling site; (b) Profile
characteristics.

Using Ruellan's terminology the following horizons
were distinguished (Fig. 1b): 0-15 cm, A1 horizon;
15-25 cm, cambic horizons; 25-29 cm, Bca 'encroûtement
massif' (massive crust) - the soil is a sol brun calcaire;
29-30 cm, 'pellicule rubanée' (ribbon-like thin film);
30-35 cm, 'croûte feuilletée' (laminated crust); 35-55 cm,
'encroûtement massif'; 55-115 cm, 'accumulation
discontinué' (discontinuous accumulation); 115-220 cm,
'accumulation diffuse' (diffuse accumulation). These
horizons are developed in gravel beds.
    There are two 'encroûtements massifs' horizons,

one with a pedological structure (cambic horizon of
'sol brun calcaire'), the other with a lithological
structure (gravel beds). The two are separated by a
thin, but well differentiated, complex formed by
'pellicule rubanée' and 'croûte feuilletée' horizons.
The differentiated forms are similar to those described
by Gile et al. (1965) whose K1 horizon corresponds to
the 'encroûtement massif' with the pedological structure,
K2m to the 'pellicule rubanée' and 'croûte feuilletée',
K2 to the 'encroûtement massif' with a lithological
structure, and K3 to the deeper less differentiated
accumulations.

FORMS AND OCCURRENCE
     Samples were examined in thin sections with a
petrological microscope and with a scanning electron
microscope coupled with a wave-length dispersive X-ray
analyser (SEM-WDXRA). Seven main habits and morphotypes
were identified:
1.   Sparitic crystals : pseudohexagonal in cross section,
     60-80 μm diam., often in association with plant
     material which they pseudomorphose. They occur
     discretely in voids or within aggregates and other
     accumulations. They become pinkish red when stained
     with Red Alizarine S (Dickson, 1965; Wieder and
     Yaalon, 1974; Conway and Jenkins, 1978) indicating
     calcite. They are particularly abundant in the
     'encroûtement massif' with pedological structure
     (Fig. 2 a) but also occur as deep as 200 cm (Fig.
     3 a).
2.   Micritic crystals : formed of calcite, they are
     mainly 1-3 μm in diam., pseudomorphosing root cells.
     They are easily visible under the scanning electron
     microscope particularly in the 'pellicule rubanée'
     and the 'croûte feuilletée' horizons (Figs. 2 b, 4 a
     and b).
3.   Needles : three types of needles have been identified :
     a) lath-shaped : these are smooth at low magnification
     appearing in cross-section as laths in bundles of 2
     to 4 or as rectangular features depending on the
     orientation of the sample. They are 30-40 μm long
     by 1 to 4 μm wide. They occur at all levels in the
     section. They may cross and coalesce at certain

640   J. Ducloux and P. Butel

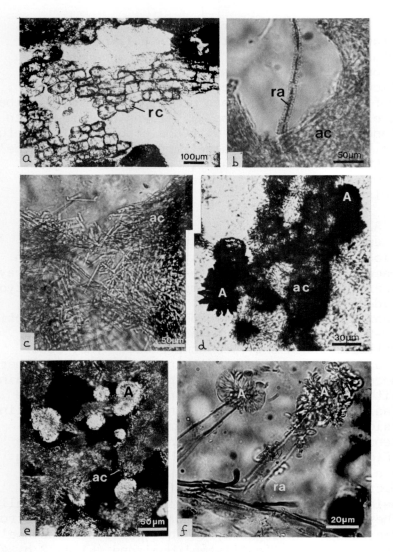

Fig. 2. Optical microscope: (a) Root with calcified
cells (rc). PPL; (b) Radicle with calcified cells
(ra) upon needles (ac). PPL; (c) Matted needles
(ac). PPL; (d) Spiny aragonite (A), coloured in
black by Fiegl's solution, with needles (ac).
PPL; (e) Aragonite crystals (A) with needles
(ac). XPL; (f) Aragonite crystals (A) upon
radicles (ra). PPL.

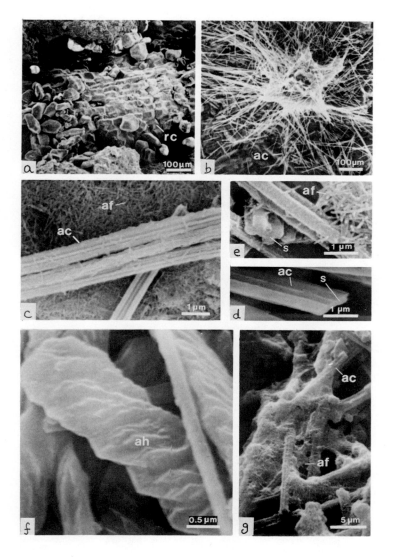

Fig.3. SEM: (a) Root with calcified cells (rc) in the
    surface horizons (Bca); (b) Needles (ac) with
    coalescence points; (c) Lath-like needles (ac)
    with very small needles (af); (d and e) Lath-
    like needle (ac) sections (s). Very small
    needles (af); (f) Herring-bone needles (ah) and
    lath-like needles (ac); (g) Lath-like needles
    (ac) cemented by very small needles (af).

J. Ducloux and P. Butel

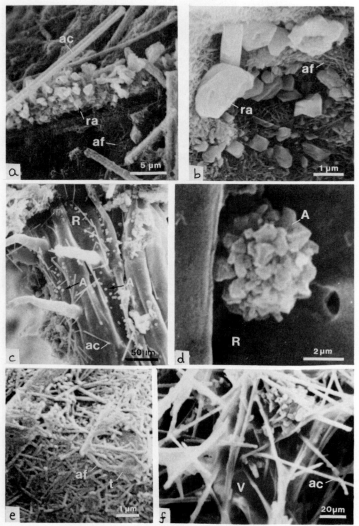

Fig. 4. SEM: (a) Calcified radicle (ra) with lath-
like needles (ac) and very small needles (af);
(b) Small radicle crystals alone (ra), upon very
small needles (af); (c) Small aragonite crystals
(A) with lath-like needles (ac) upon a root (R);
(d) Aragonite crystals upon a root (R). High
magnification; (e) Very small needles (af) and
cement (t); (f) Veils (V) stretching between
lath-like needles (ac).

points (Figs. 2 c, 3 b-g).
  b) smooth needles: they are observed in 'encroûtements massifs' and 'croûte feuilletée' horizons
where they are 0.5 μm long by 0.1 μm wide. They
form a thick matted texture with numerous coalescence
points and often combine with, or overlie, lath-
shaped needles (Figs. 3 c and 4 c).
  c) twisted needles: these often form a herringbone
pattern. The needles are 20 μm long by 1 to 8 μm
wide. Stoops (1976) described this type as
lublinite following the terminology of Polish
authors. They appear to consist of piles of small
hexagonal crystals. They have been noted among
lath-shaped crystals (Fig. 3 f).
4. Micritic cement: constituent crystals are much < 1 μm
diam. The cement, which appears to form by coalescence
of small smooth needles (Fig. 4 e), is best developed
in the 'pellicule rubanée' horizon but also occurs
discontinuously in the 'encroûtements massifs' and
the 'croûte feuilletée' horizon.
5. Fibrous associations: they have a spiny form (resem-
bling a sea urchin) or are fan-like. Their maximum
diameter is 30 μm. They occur in voids in the
'encroûtements massifs' and 'croûte feuilletée'
horizons. They occupy three types of sites: (a)
coatings around embedded soil peds; (b) in combination with mats of calcite needles; and (c) on organic
material. They assume a black colour when stained
with Fiegl's solution suggesting that they are formed
of aragonite (Fig. 2 d, e and f).
6. Sphaerules: these vary in size from a few μm to more
than 50 μm. They appear to be formed of aragonite
(Fig. 4 c and d).
7. Veils: this form stretches between lath-shaped
needles and is found in the 'accumulation discontinué'
and diffuse horizons (Fig. 4 f). They are perhaps
amorphous carbonates.

CONCLUSIONS
    Seven main types of carbonate features have been
identified at Vouillé which can be classed into five
groups:

1. Sparitic and micritic grains of calcite arising
    from epigenesis of roots and root cells. They are
    authigenic and not detrital like those observed by
    Al Rawi et al. (1968) and Altaie et al. (1969) in
    arid areas.
2. Needles of various size which have been described in
    a number of soils (Kubiena, 1970; FitzPatrick,
    1971; Nahon, 1976). Such needles have been termed
    lublinite by Sehgal and Stoops, 1972: X-ray analysis
    has confirmed their calcitic nature.
3. Cement, common in 'pellicule rubanée' horizons, is
    responsible for gradual thickening of horizons.
    The uppermost layer generally represents the youngest
    cement (Vermeire et al., 1974). The cement appears to
    derive from the coalescence of needles.
4. The origin of the spiny and fan-like forms of aragonite
    is still not fully understood. Organic matter appears
    to play some part in their formation. Such forms
    have been described in Morocco as well as in western
    France (Nahon et al., 1980).
5. The final group consists of veils. Their calcitic
    nature has not been absolutely demonstrated. They
    appear to be rather rare. They may be silico-alumina
    amorphous gels.
    With respect to the relationship between morphology
of the carbonate concentrations and their mineralogical
nature and particular horizons, the following observations
can be made : (a) the 'encroûtements massifs' horizons
appear to be characterised by aragonite which is not found
in other horizons in the section; (b) the 'pellicule
rubanée' horizon seems to be made up almost exclusively of
cement; (c) the 'croûte feuilletée' horizon seems to be
composed of practically all types of crystallisations but
particularly cement and matted development of needles
in voids; (d) the 'accumulations discontinués et
diffuse' mostly contain lath and twisted needles and
some veils.
    Finally, with respect to terminology, that of
Ruellan, which is well adapted to description of calcareous
soil profiles, also links well with the description of the
mineralogical aspects.

REFERENCES

Al Rawi, G.J., Sys, C. and Laruelle, J. 1968. Pedo-
    genetic evolution of the soil of the Mesopotamian
    Flood-Plain. Pédologie, 18, 63-109.

Altaie, F.H., Sys, C. and Stoops, G. 1969. Soil groups of
    Iraq. Their classification and characterization.
    Pédologie, 19, 65-148.

Conway, J. and Jenkins, D. 1978. Application of acetate
    peels and microchemical staining to soil micro-
    morphology. In: M. Delgado (Ed), Micromorfologia
    de Suelos. Universidad de Granada, 47-58.

Dickson, J.A.D. 1965. A modified staining technique for
    carbonates in thin section. Nature, 205, 587.

FitzPatrick, E.A. 1971. Pedology. A Systematic Approach
    to Soil Science. Oliver and Boyd, Edinburgh,
    306 pp.

Gile, L.H. 1961. A classification of Ca-horizons in soils
    of a desert region, Dona Ana County. New Mexico.
    Soil Sci. Soc. Amer. Proc., 25, 52-61.

Gile, L.H., Peterson, F.F. and Grossman, R.B. 1965. The
    K-horizon : a master soil horizon of carbonate
    accumulation. Soil Sci., 99, 74-92.

Kubiena, W.L. 1970. Micromorphological Features of Soil
    Geography. Rutgers Univ. Press, New Jersey, 254 pp.

Nahon, D. 1976. Cuirasses ferrugineuses et encroûte-
    ments calcaires au Sénégal occidental et en
    Mauritanie. Systèmes évolutifs, géochimie, struc-
    tures, relais et coexistence. Sciences Géologiques,
    Strasbourg, 245 pp.

Nahon, D., Ducloux, J., Butel, P., Augas, C. and Paquet, H.
    1980. Néoformation à aragonite dans les encroûte-
    ments calcaires. Comptes Rendus de l'Academie des
    Sciences, Paris, Série D, 291, 725-727.

Ruellan, A. 1970. Les sols à profil calcaire différencié
    des plaines de la basse Moulouya (Maroc oriental).
    Thèse, Univ. Strasbourg, 482 pp.

Sehgal, J.L. and Stoops, G. 1972. Pedogenic calcite
    accumulation in arid and semi-arid regions of the
    Indo-Gangetic alluvial plain of erstwhile Punjab
    (India). Their morphology and origin. Geoderma,
    8, 59-72.

Stoops, G. 1976.   On the nature of "lublinite" from
     Hollanta (Turkey).   Amer. Mineral., 61, 172.
Vermeire, R., Dauchot-Dehon, M. and Paepe P. de,1974.
     Sur l'âge d'une croûte calcaire de la zone
     occidentale de l'île de Euertepentura (Canaries).
     Pédologie, 24, 40-48.
Wieder, M. and Yaalon, D.H. 1974.   Effect of matrix
     composition on carbonate nodule crystallization.
     Geoderma, 11, 95-121.

THE MICROMORPHOLOGY OF SPODOSOLS IN CATENARY SEQUENCES
ON LOWLAND HEATHLANDS IN SURREY, ENGLAND

R.I. Macphail

Department of Human Environment, Institute of Archaeology,
University of London

ABSTRACT
      A model accounting for two microfabric types in
spodic horizons has been developed by De Coninck. A
study of the chemistry and micromorphology of catenary
sequences on lowland heaths in Surrey suggests that the
type of Spodosol microfabric is related to slope-unit
position, because profile development is governed by the
prevalence of vertical or lateral movement of illuvial
materials. These effects give rise to a greater
variety of spodic fabric types than suggested by De
Coninck.

INTRODUCTION
      Much interest has been shown in the micromorphology
of Spodosols as a means of understanding their pedogenesis. One model advanced by De Coninck (1980) accounts
for the two main microfabrics present in spodic horizons,
i.e. loose horizons of polymorphic pellets and aggregates,
mainly arising from biological influences, whereas cemented
horizons of monomorphic coatings develop when immobilisation of organo-metallic compounds predominates. This
scheme, when considered in relation to catenary sequences
on lowland heaths in Surrey, is found to be complicated
by the probable movement of organic matter and organically
mobilised sesquioxides downslope. This process gives
rise to predictable soil development-slope unit associations, which can be understood by examining the microfabrics of the Spodosol phases identified. In a previous
study (Macphail, 1979), Surrey heathland soils were
mapped at the phase level, samples from representative
soil types analysed for iron and carbon, and samples
from particular horizons described micromorphologically.
A model was developed which ascribed specific soil types
and processes to plateau and slope positions. This

647

paper selects micromorphological, iron and carbon data from Blackheath (south-east of Guildford) and Headley Heath (north-east of Dorking) to briefly illustrate some aspects of these soil-slope associations.

MATERIALS AND METHODS

The soils from Blackheath are developed in medium sandy Folkestone Beds of Cretaceous age and those from Headley Heath in medium sandy Headley Sands of supposed Calabrian age. The following measurements of pyro-phosphate extractable carbon and iron were made according to the method of Bascomb (1968): C in extract (C_{ext}); soluble Fe not precipitated by ammonia (Fe_{sol}); colloidal Fe precipiated by ammonia (Fe_{ppt}); and residual Fe in dithionite extract (Fe_{res}). Some 1000-1100 points were counted on thin sections (mainly 75 x 35 mm) for fabric analysis and the following identified: (1) organic matter (OM); (2) organo-sesquioxidic pellets (OSP) - a type of polymorphic material; (3) sesquioxidic pellets (SP) - a type of polymorphic material; (4) organo-sesquans (OS), after Brewer (1964) - monomorphic coatings; (5) sesquans (S) - monomorphic coatings; and (6) amorphous sesquioxidic concretions (ASC). Mineral grains, voids etc. were also counted but results are not reported here.

RESULTS

The model divides Spodosols into phases (Table 1). Types P1 and P2 in plateau positions exhibit strong eluviation and a continuous Bh horizon (Munsell value < 3, chroma < 2) over a Bhs horizon (Avery, 1973) (Munsell value $\leqslant 4$, chroma generally $\leqslant 6$). In contrast, types S1, S2 and S3 on sloping sites have discontinuous Bh horizons merging with Bhs material. Type S5 in the 'slope' group is a special case in having a diffuse boundary between Ea and Bs horizons. On Blackheath types P1, S2, S3 and S4 were examined and on Headley Heath types P1, S3, S4 and S5 from one catena.

Bh(f) and Bhs horizons tend to contain more organo-sesquioxidic pellets and organo-sesquans than Bs horizons but less sesquans and amorphous sesquioxidic

Table 1.   Spodosol phases.

Type   :   ORTHOD

Plateau

P1   Horizontal continuous Bh horizon
P2   Tongued continuous Bh horizon

Slope

S1   Horizontal discontinuous Bh horizon
S2   Tongued discontinuous Bh horizon
S3   Discontinuous Bh horizon, Bs
                mottles within Bhs horizon
S4   No Bh, only Bhs horizon

Type   :   FERROD

S5   No Bh or Bhs horizon, only a Bs
                horizon

concretions (Table 2). Ea2 horizons in slope soils
may also contain large amounts of organo-sesquioxidic
and sesquioxidic pellets (Samples 3 and 14). Larger
amounts of iron were extracted by pyrophosphate from
Bh and Bhs horizons than from Bs horizons.

DISCUSSION
     Type P Spodosols are characterised by clear Ea
horizons, a sharp eluvial-illuvial horizon boundary
and a continuous Bh horizon with relatively large
amounts of organo-sesquioxide pellets, organo-sesquans
and pyrophosphate extractable iron. In the Ea horizon
the mineral grains are generally uncoated. The B
horizon grains are more or less coated and agglutinated
by polymorphic material. The profile features are
consistent with the vertical migration of organic
matter and organically-complexed sesquioxides and their
illuviation in the spodic horizon (Fig. 1a). By
contrast, the Ea horizon of slope soils contains varying

Table 2. Micromorphological and Chemical Data (%).

Sample number	Soil type	Horizon	OM	OSP	SP	OS	S	ASC	C_{ext}	Fe_{sol}	Fe_{ppt}	Fe_{res}	Fe_{ext}
BLACKHEATH Plateau													
1.	P1	Bh(f)2	0.3	27.0	9.3	6.4	0.2	0.2	1.32	0.30	0.34	0.98	0.64
Slope													
2.	S2	Ea2	2.6	39.0	0.9	3.9	1.4	0.4	1.65	0.0	0.12	0.88	0.12
3.		Ea2	-	2.8	28.0	-	2.0	1.8	2.07	0.25	0.27	0.56	0.57
4.	S3	Bhs	-	13.6	16.0	2.0	8.7	1.2	1.43	0.16	0.30	0.74	0.46
5.		Bs(mot.)	-	5.9	21.0	0.7	4.5	1.3	1.41	0.0	0.02	1.20	0.02
6.	S2	Bs	-	0.1	9.0	-	17.0	10.0	N.D.	N.D.	N.D.	N.D.	N.D.
7.	S4	Ea2	-	5.1	2.7	0.5	3.5	-	N.D.	N.D.	N.D.	N.D.	N.D.
8.	S4	BHs	1.4	11.5	0.2	8.7	1.2	-	N.D.	N.D.	N.D.	N.D.	N.D.
9.	S4	Bs	-	0.2	0.4	-	9.5	1.8	N.D.	N.D.	N.D.	N.D.	N.D.
HEADLEY HEATH Plateau													
10.	P1	Ea2	31.0	-	-	-	-	-	0.26	0.01	0.02	0.0	0.03
11.	P1	Bhs	-	3.7	11.6	5.7	17.5	22.0	0.77	0.17	0.10	0.03	0.26
Slope													
12.	S2	Eah	15.6	2.5	-	1.6	-	-	0.31	0.02	0.01	0.03	0.03
13.	S4	Bhs	0.1	3.5	13.6	1.0	12.8	0.3	0.68	0.16	0.10	0.06	0.25
14.	S5	Ea	9.3	-	17.0	-	0.7	1.1	1.79	0.09	0.17	0.15	0.26

N.D. : No Determination

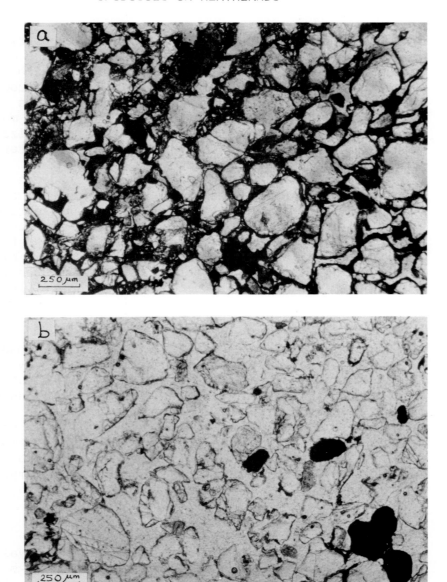

Fig. 1.(a) Strongly cemented Bs horizon of soil
     type P1, XPL; (b) Weakly cemented Bs horizon
     of soil type S4, PPL.

amounts of polymorphic material, some of which is rich
in organic matter. In most cases, the Ea horizons contain
patches of underlying spodic horizon material which occur
in one of the following forms: (1) patchy Bh horizon
material with mainly polymorphic material; (2) Bhs
horizon material containing patches of polymorphic material
low in organic matter (Bs mottles); (3) a Bhs horizon
material with only small amounts of polymorphic material,
and a weakly cemented Bs horizon (Fig. 1b).
        There is both a vertical and lateral component to
water movement in slope positions and the features of
profiles on slope sites are thought to be a reflection
of the lateral flow component in particular. Thus,
whereas the Ea horizons in plateau profiles appear to
result from loss of constituents,these horizons in slope
profiles may be subject to increments and/or losses of
material. Accumulation of sesquioxides in receiving
sites is believed to cause reduction in complexing
agent/sesquioxide ratio, making sesquioxides more
insoluble and causing immobilisation of the iron
(Schnitzer and Skinner, 1964; McKeague and St. Arnaud,
1969). Immobilised sesquoixides are represented by
large numbers of sesquioxidic pellets. The spodic horizons
of slope profiles appear also to be affected by losses of
material as well as gain.
        The fabric of spodic horizons and of Ea horizons
from slope profiles is thus complex. Polymorphic fabrics,
particularly of the sesquioxidic type, appear to be more
related to slope movement of material than to a purely
faunal origin.

ACKNOWLEDGEMENT
        The author wishes to acknowledge the technical
assistance of K.S. Janes and the support of the Royal
Borough of Kingston-upon-Thames (Surrey) and thanks
H.C.M. Keeley for reading and C. Skelton for typing the
manuscript.

REFERENCES
Avery, B.W. 1973. Soil classification in the Soil Survey
    of England and Wales. J. Soil Sci., 24, 324-338.
Bascomb, C.L. 1968.   Distribution of pyrophosphate-
    extractable iron and organic carbon in soils of
    various groups. J. Soil Sci., 19, 251-268.
Brewer, R. 1964. Fabric and Mineral Analysis of Soils.
    John Wiley and Sons, New York, 470 pp.
De Coninck, F. 1980. Major mechanisms in formation of
    spodic horizons. Geoderma, 24, 101-128.
Macphail, R.I. 1979. Soil variation on selected Surrey
    heaths. Ph.D. Thesis, C.N.A.A., Kingston
    Polytechnic.
McKeague, J.A. and St. Arnaud, R.J. 1969. Pedotrans-
    location : eluviation-illuviation in soils during
    the Quaternary. Soil Sci., 107, 428-434.
Schnitzer, M. and Skinner, S.I.M. 1964. Organo-metallic
    reactions in soils : 3. Properties of iron- and
    aluminium organic-matter complexes, prepared in the
    laboratory and extracted from a soil. Soil Sci.,
    98, 197-203.

MICRO-ORGANISATION OF LOOSE FERRALLITIC MATERIALS IN THE CAMEROONS

J.-P. Muller

Office de la Recherche Scientifique et Technique
Outre-Mer, 70-74 Route d'Aulnay, 93140 BONDY, France
(DGRST, Institut de la Recherche Agronomique, Centre
Agronomique de Nkolbisson, BP 2017, Yaoundé, Cameroon)

ABSTRACT
     In the Cameroons, loose clay horizons which occur on
the relict indurated formations are often several metres
thick. They are generally homogeneous from the geochemical, mineralogical and physico-chemical points of
view and are distinguishable from one another mainly by
their colour and their organisation. The vertical sequence
of organisation is the morphological expression of two
opposing pedogenic processes :
     - A process of organisation involving microstructura-
tion by reticulation of the original clay plasma, which
affects the deep, red and dense horizons.
     - A process of disorganisation resulting from
discolouration, plasmic microlysis (or 'crumbling' of the
microstructures) and disturbance of the s-matrix (with
clay eluviation). These three mechanisms, acting either
together or successively, affect the upper part of the
profiles.
     The relative development of these two processes is
variable. Moreover, it is not possible to define a
strict chronology since the weak state of organisation
of the upper horizons could be inherited partly from the
original pedoplasmation.

INTRODUCTION
     About two-thirds of the Cameroons is covered by
ferrallitic soils with a long and complex history
(Fig. 1). Unlike many other African countries, particularly those in western Africa (d'Hoore, 1954), except
in rejuvenated zones (Segalen, 1967; Martin, 1967,
1970; Muller, 1979), relict indurated formations in
the Cameroons generally occur below a loose clay formation

J.-P. Muller

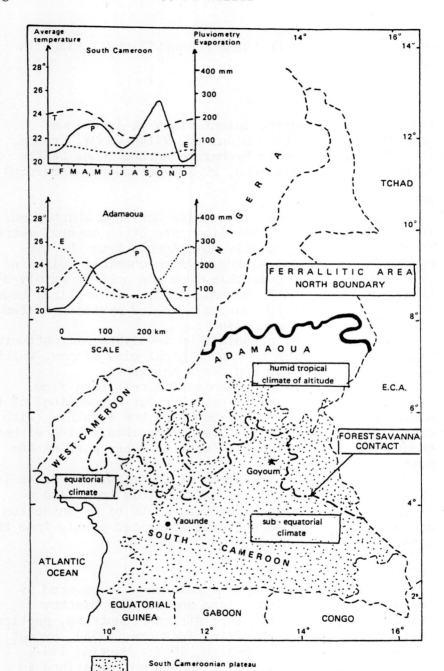

Fig. 1. Location map.

several metres thick (Muller, 1975).

From a geochemical, mineralogical and physico-chemical point of view, this loose clay formation is fairly homogeneous over the south Cameroons plateau (Morin, 1979). The clay fraction is kaolinitic; goethite, hematite and,to a lesser extent,gibbsite are the main hydroxides and oxides; the skeleton grains are almost exclusively quartzose; primary minerals, other than some heavy minerals, are entirely transformed; the horizons are strongly leached of exchangeable bases (strongly unsaturated soils (C.P.C.S., 1967)).

The macroscopic study of this loose formation has been based previously on observation of shallow profiles (generally less than 2m deep). Because of the weak contrast in structure between horizons and the relative uniformity of texture throughout, classification of the soils at intermediate taxonomic levels (suborders) has been based on colour ('red', 'grey ferruginous' and 'yellow' soils).

Observations made on sections (Bocquier and Muller, 1973) or deep profiles reveal a more or less pronounced vertical colour gradient from lower to upper horizons (Fig. 2). Moreover, detailed examination, e.g. with a microscope, shows typical horizons and volumes and the existence of specific structural organisations. The macro- and micromorphological components of this loose layer are examined in this paper.

VERTICAL ORGANISATION IN THE LOOSE LAYER

Five horizons are recognised in the material, divided into a lower and an upper group (for full descriptions, see Muller, 1977a and b).

1. The group of lower horizons is characterised by weakly contrasted colours and diffuse boundaries (Fig. 2a). It includes from the base of the section upwards:

(a) Deep dense horizons

Macromorphology: These are the reddest horizons within the layer with Munsell hues of 10R. Other properties include clay texture, moderately developed coarse blocky structure, compact and firm consistency

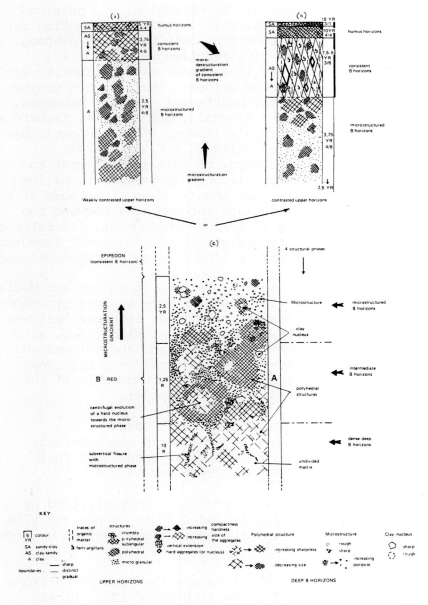

Fig. 2. Diagram of the vertical evolution of
         loose horizons.

and an apparent low porosity.

Micromorphology: Few randomly distributed quartz
grains; weakly oriented argillasepic plasmic fabric
with a few insepic patches (Fig. 3a), gradually becoming
insepic upwards with heterogeneous masepic patches;
some plasma separations surrounding and delimiting
micropeds (50-200μm) between crossed polarisers;
microporosity composed of micro-fissures and scattered
meso- and orthovughs; few aggrotubules.

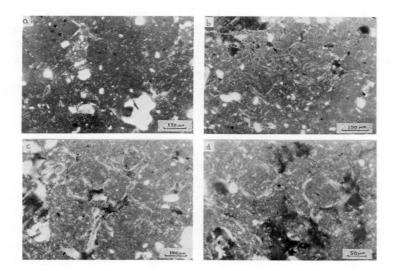

Fig. 3(a) Red, dense deep B horizon. Weakly insepic
     plasma. Isolated microvoids; (b) Horizon
     including dense and microstructured phases.
     Strongly expressed plasmic fabric (in-lattisepic)
     bordering a void (top left). Strongly bire-
     fringent plasma separations. Differentiated
     micropeds are contiguous on a local basis. Some
     small peripheral fissures. Structured and
     darker plasma lower right; (c) Horizon similar
     to (b). Contrasted micropeds partly or totally
     separated by micro-fissures which anastomose in
     the form of a star; (d) Microstructured
     horizon. Interpedal micro-fissures evolve
     into a granular fabric.

(b) Horizons with mixed dense and microstructured areas
     Macromorphology: Dense red phase (c. 10R) continuous
with underlying dense horizon; more organised than underlying dense horizon; finer but still compact structure;
some dense areas of very low porosity; gradual boundary
to a blocky phase with fine to very fine peds; less red
in colour (2.5YR); more or less loose structure; very
porous; friable to very friable.
     Micromorphology: The dense phase has a mainly
insepic, and the granular phase a ma-in-lattisepic,
plasmic fabric (Fig. 3b); micropeds, more numerous than
in the horizon below, are also more discrete; microped
tend to be more contrasted with the surrounding matrix
due to the surrounding plasma separation being lighter
coloured and more birefringent than the matrix; near
the top of the horizon the microstructuration becomes
more developed; microped are partly or totally
separated by microfissures which anastomose in the shape
of a star (Fig. 3c).

(c) Microstructured horizons
     Macromorphology: Red to yellowish red (2.5YR-
5YR); strongly developed microgranular structure and
large porosity; very friable; little dense matrix,
occurring as rounded bodies (clay nucleii) a few μm to a
cm in diameter.
     Micromorphology: Skeleton similar to deeper horizons;
more strongly expressed plasmic fabric (in-vo-lattisepic);
dense network of fissures increasing upwards to develop
a reticulate pattern (Fig. 3d); abundant rounded microped
with sharper contrast than in horizons below due to
further discolouration of surrounding plasma separations;
strong pedoturbation; few clay nucleii (Fig. 4a), with
weakly insepic or asepic plasmic fabric.

2. The group of more or less contrasted horizons
affected by organic accumulation shows a diffuse or gradual
transition to the deeper horizons described above (Fig. 2b
and c). They include:

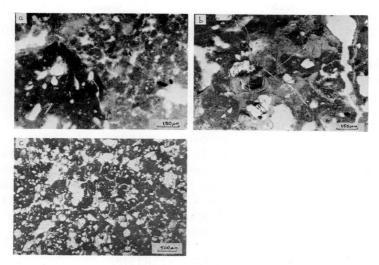

Fig. 4. (a) Microstructured horizon. Areas of
        apedal and isotic plasma are separated in
        'clay nucleii' (left) within a strongly
        microstructured matrix subjected to pedo-
        turbation (right); (b) 'Consistent' horizon.
        Compact with a weakly insepic plasmic fabric.
        Composed of long fissures. Ferri-argillans
        with simple or zoned voids; (c) Humus horizon.
        Abundant quartzose skeleton grains. Quasi-
        isotic structure. Irregular and inter-
        connected meso- and macro-vughs.

(a) 'Consistent' B horizons
      Macromorphology: More or less darkened by organic
matter throughout and/or on the faces of aggregates;
red to yellow plasma (2.5YR to 7.5YR); clayey to sandy
clay; medium to coarse compact blocky structure;
moderate porosity, distinct to sharp upper boundary.
      Micromorphology: Degree of development of the plasmic
fabric decreases upwards from ma-insepic to weakly insepic
to locally isotic in the upper part and in lighter
pockets; fewer micropeds than in underlying horizon and
less discrete; some patches of redder plasma decreasing
upwards in number and size; porosity mainly in the form
of ortho- macro- and mesovughs and some vertically

oriented ortho-fissures;   simple and complex continuous
and zoned ferri-argillans and organo-argillans along
microfissures, thickest in vertical fissures (Fig. 4b).

(b)   Humus horizons
        Macromorphology: Yellowish red to dark yellow (5YR to
10YR) stained by organic matter;   clayey sand to sandy
clay;   subangular to nodular blocky structure;   loose;
porous;   very friable.
        Micromorphology: Abundant quartzose skeleton;
weakly insepic plasmic fabric becoming asepic in yellower
parts;   no micropeds;   agglomeroplasmic   related distribution;   many ortho- meso- and macrovughs strongly interconnected by ortho-fissures (Fig. 4c);   number and size
of organo- and ferri-argillans decrease upwards;   no
argillans in the uppermost part.

MECHANISMS RESPONSIBLE FOR ORGANISATION
        Apart from the gradients in colour and texture, which
are generally weak, the horizons are distinguished mainly
by their structure.   The continuous vertical study
indicates two main mechanisms:

Mechanisms of organisation
        In the lower part, the gradual evolution from red
compact, firm, rather impervious phases to very porous,
very friable, microgranular phases is associated with
the gradual development of a blocky structure in which
the discreteness of the units increases and size decreases
upwards.
        This fractionation which is visible macroscopically
as well as microscopically, corresponds to microstructuration and the genesis of microped organisations.
It begins with a simple change in plasmic fabric without
change in colour, followed by further reticulation of the
matrix and modification of the original plasmic fabric
as fractionation proceeds.
        This structuration becomes more marked upwards,
developing ultimately through peripheral microfissures
into a granular fabric.   Part of the matrix remains
resistant to microstructuration and residual clay nucleii,
corresponding to original B horizon material, remain even

near the surface.

Mechanisms of disorganisation
     Three phenomena are observed in the upper part of
the profiles:
     (a) Discolouration of the plasma (i.e. change from
reddish to yellowish colour): The first signs of colour
change observed in thin sections are in samples from deep
in the profile. The change is noted in fine lamina in
the weakly structured soil material. Macroscopically,
the change in colour is only clearly visible in the upper
part of the profile. The observed colour gradient varies
in intensity and extent (Fig. 5). The change in colour is
probably related to decomplexing within the original
plasma (Segalen, 1969).

sharp boundary

1. Red and compact phase 3,75 YR-2,5 YR. Sharp fine and polyhedral structure.

2. Discolored and very compact phase 5, YR-7,5 YR. Not very sharp, polyhedral, fine/mean structure.

3. Microstructured phase 5 YR, very loose arrangement subject to strong pedoturbation.

a - horizon 150-210 cm

2 cm

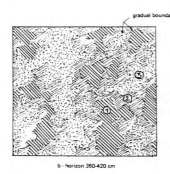

gradual boundaries

1. Red and compact phase 2,5 YR-10 R. Sharp/very sharp, polyhedral and fine structure.

2. Red phase 2,5 YR-3,75 YR. Sharp polyhedral and very fine structure under a microgranular.

3. Microstructured phase 3,75 YR. Loose arrangement on a localized and very fine polyhedral structure.

b - horizon 360-420 cm

Fig. 5. Structural phases in two horizons of a red
        soil which is discoloured in its upper part
        (profile GOY344).

    (b) Concentration of the plasma in the consistent
horizons:
This affects all the ferrallitic soils but it is more
intense and widespread the more marked the change in
colour from red to yellow. It is possible, especially
in soils that are red to the surface, that this concentration is partly relict, i.e. inherited from
previously deeper red B horizons. However, it also seems
to characterise situations in which there is fusion of
microstructure together with discolouration (microlysis:
Muller, 1977b). Associated with discrete and localised
discolouration is an attenuation of the birefringence of
plasma separations which becomes more marked towards
the surface.
    (c) Instability of the soil material: The material
becomes more unstable in the yellower 'microlysed' upper
horizons. There is dispersion of the plasma which is
translocated into the 'consistent' B horizons leaving
a relative accumulation of quartzose skeleton in the
humus horizons.
    There is thus a clear correlation between the
evolution of the morphological components - colour,
structure and texture. Increasing colour change from
redder to yellower colours is associated with more
intense microlysis which facilitates translocation of
clay.

CONCLUSIONS
    The differentiation of clayey material in many
ferrallitic soils in the Cameroons is highly dependent
on the relative development of two pedogenic processes
which develop from lower toward upper parts of a profile.
One is a factor of organisation, the other one of
disorganisation.
    There is an evident connection between the phenomena
of discolouration, plasmic microlysis and leaching,
characteristic of the disorganisation process in the upper
horizons.
    It is difficult to establish a strict chronology for
the two processes. Traditionally,a continuous vertical
connection is inferred particularly when the transition
between horizons is gradual, each horizon being assumed
to form by direct transformation of the underlying

horizon. Observations above suggest that the vertical
sequence of organisation is not systematically linked to
a particular sequence of pedological events, i.e. microstructuration followed by microlysis. Compact red
original material with poorly developed plasmic fabric
(argillasepic, insepic) can occur in microstructurally
developed horizons even near the surface in a state
close to the original.

   Nevertheless the two processes oppose each other.
One is an agent of stability producing structural units
resistant to degradation. The other destroys the initial
structure and eventually causes disturbance of the soil
material.

REFERENCES
Bocquier, G. and Muller, J.-P. 1973. Les coupes du
     chemin de fer Transcamerounais de BELABO à
     NGAOUNDERE. Reconnaissance pédologique. Centre
     ORSTOM de Yaoundé, multigr., 29pp.
C.P.C.S. 1967. Classification Française des Sols. Ecole
     Nat. Sup. Agr. Grignon, (Mimeographed), 87 pp.
D'Hoore, J. 1954. La carte des sols d'Afrique au 1/5,
     000,000. Notice explicative. Publ. 33, C.C.T.A.
     Lagos, 209 pp.
Martin, D. 1967. Géomorphologie et sols ferrallitiques
     dans le Centre Cameroun. Cah. ORSTOM, sér. Pédol.,
     5, 189-218.
Martin, D. 1970. Quelques aspects des zones de passage
     entre les surfaces d'aplanissement (Centre Cameroun).
     Cah. ORSTOM, sér. Pédol., 8, 219-241.
Morin, S. 1979. Géomorphologie du Cameroun. Atlas
     Jeune Afrique du Cameroun.
Muller, J.-P. 1975. Quelques réflexions sur les ter-
     minologies, taxonomies et méthodologies appliquées
     aux sols des domaines ferrallitiques de Côte
     d'Ivoire et du Cameroun. In: C.R. d'une Tournée
     dans le Nord de la Côte d'Ivoire. Centre ORSTOM
     Abidjan, 18-22.
Muller, J.-P. 1977a.   Microstructuration des struc-
     tichrons rouges ferrallitiques, à l'amont des
     modéles convexes (Centre Cameroun). Aspects

    morphologiques.   Cah. ORSTOM, sér. Pédol., 15,
        25-44.
Muller, J.P. 1977b.   La microlyse plasmique et la
        différenciation des épipédons dans les sols ferr-
        allitiques rouges du Centre Cameroun.   Cah.
        ORSTOM, sér. Pédol., 15, 345-359.
Muller, J.P. 1979.   Carte des sols du Cameroun à
        1/3,250,000.   Atlas Jeune Afrique du Cameroun.
Segalen, P. 1967.   Les sols et la géomorphologie du
        Cameroun.   Cah. ORSTOM. sér. Pédol., 5, 137-187.
Segalen, P. 1969.   Contribution à la connaissance de
        la couleur des sols à sesquioxydes de la zone
        intertropicale : sols jaunes et sols rouges.
        Cah. ORSTOM, sér. Pédol., 7, 225-236.

MICROMORPHOLOGY OF SOME LATERITIC SOILS IN MALAYSIA

S. Zauyah

Soil Science Department, Universiti Pertanian Malaysia,
Serdang, Selangor, Malaysia

ABSTRACT
     Lateritic soils in Malaysia are either formed in
situ or from colluvial material. The main micromorphological criterion used to identify these two
groups of soils is the type of lateritic nodules
present. Soils that formed in situ consist of
irregularly-shaped sesquioxidic nodules or platy
laterised parent material. The petroplinthites have
some void neo-ferrans, angular quartz grains and patches
of crystalline goethite. The platy laterised parent
material,occurring in the lower horizons,have droplets
of iron oxides in the matrix, bands of crystalline
goethite and void neo-ferrans. The lateritic soils
which have well rounded sesquioxidic nodules containing
runiquartz, irregular bands of goethite, void goethans
and neo-ferrans are believed to be formed from colluvial
material. SEM studies of these nodules show a network
of tubular goethite surrounded by a matrix of kaolinite.
The plasmic fabric of all the soils range from isotic to
insepic in the oxic horizons to vo-omnisepic in the
argillic horizons. The microstructure of the oxic
horizon is generally crumbly while that of the argillic
horizon is irregular jointed.

INTRODUCTION
     Lateritic soils have been mapped in many parts of
Malaysia. For purposes of mapping and correlation,
these soils are now defined as having a thick (>25 cm)
band of lateritic gravels with its upper boundary
within 1 m of the soil surface (Paramanathan, 1980).
These soils have been found to be either developed in
situ mainly over shale, siltstone, phyllite and schist,
or from colluvial material deposited on an old erosional
surface.
     Micromorphological studies have been made on one

of the major lateritic series (Malacca series) studied
in detail by Eswaran and Daud (1980). The fabrics of
the plinthite and petroplinthite found in Oxisols have
also been described by Eswaran (1979). This paper
reviews the micromorphological features of some of the
other lateritic soil series in Malaysia which are now
grouped,from field characteristics,as those that
formed in situ or from colluvial material.

MATERIALS AND METHODS
    Bulk samples and undisturbed soil samples were
collected from the major horizons of twelve profiles.
All the soils contain more than 50% nodules. The
texture is generally clayey. The B horizon of three
profiles are oxic while the rest are argillic.
    The mineralogy of the clay fractions and the nodules
from the bulk samples was analysed by XRD. Thin sections
were prepared from the undisturbed soil samples. The
terminology used in the description is from Brewer (1976)
and Stoops and Jongerius (1975). Small fragments of
the lateritic nodules were selected for SEM observation
using the method outlined by Stoops (1970).

MICROMORPHOLOGY
    The micromorphological investigation of these soils
reveals that they can be separated into two main groups
using the characteristics of the sesquioxidic nodules
as the main criteria.

    Soils with Rounded Nodules
    The B horizons of these soils are either argillic
or oxic. The Oxisols exhibit a crumbly microstructure
with the s-matrix dotted with iron oxyhydrates (10 μm).
The plasmic fabric is isotic to insepic. The
pedological features are mainly the well-rounded
sesquioxidic nodules which have very well-defined
boundaries (Fig. la). These nodules contain quartz
grains which have iron oxides in the cracks (termed
runiquartz by Eswaran et al., 1975). The fabric
surrounding these grains has void neo-ferrans and
crystalline goethans. Observations of natural surfaces
of these nodules by SEM reveal a network of tubular

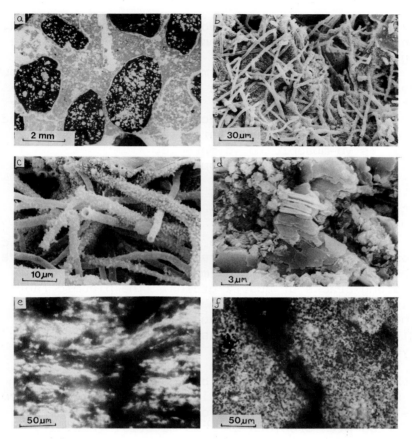

Fig. 1(a) Well rounded sesquioxidic nodules with
well defined boundaries in an Oxisol; (b) SEM
micrograph of natural surface of a rounded
nodule showing a network of tubular goethite;
(c)SEM micrograph of tubular goethite with rough
surface; (d) SEM micrograph of natural surface
of a nodule showing platelets of kaolinite;
(e) Iron oxides along schistosity plane of a
laterised schist fragment; (f) Droplets of iron
oxides surrounding neo-ferran in the matrix of a
laterised shale fragment.

goethite (Fig. 1b and c). These goethites correspond
to the yellow patches seen under the binocular microscope.
Such tubular goethite was also reported by Comerma et al.

1977) in their study of petroplinthite from Venezuela.
Other parts of the fabric show platelets of kaolinite
(Fig. 1d).
    The Ultisols contain similar types of nodules, but
occasionally skeletal grains are absent. Crystalline
goethite occurs in irregular bands along the crust or
as void goethans. Other pedological features observed
are channel and vugh ferri-argillans. The plasmic
fabric is vo-mosepic to vo-masepic.
    The c/f related distribution in all the profiles
is close porphyric.

   Soils with Irregular Shaped Nodules or Laterised
   Parent Material
    The B horizons of these soils are either argillic
or oxic. The Oxisol exhibits a crumbly microstructure.
Gibbsite also occurs in the matrix as small nodules
and along cracks of quartz grains. The plasmic fabric
is isotic. The Ultisols generally have an irregular
jointed microstructure. Channel and vugh ferri-
argillans are common and plasmic fabric is vo-masepic
to vo-omnisepic.
    The laterised fragments in the lower horizons
of the soils show that iron oxides have moved along
the schistocity plane (Fig. 1e) or along voids to form
neo-ferrans (Fig. 1f). The other part of the matrix
is dotted with droplets of this material (Fig. 1f)
very similar to those described by Hamilton (1964). In
the upper horizons, bands of crystalline goethite
(Fig. 2a) in these fragments are observed and the
matrix is replaced by more iron oxide. Fragments of
shales which are completely laterised (Fig. 2b) show
small angular quartz grains in a matrix of the oxides.
The fragments under SEM show that matrix is dominated by
flakes of kaolinite which are stacked vertically (Fig.
2c) and also tubular goethite in the voids (Fig. 2d).
In some areas, the voids have a pitted appearance
(Fig. 2e) and at higher magnification a framework
structure, probably of goethite.
    The irregular-shaped nodules (petroplinthites)
occur in Ultisols which have variegated horizons
below them. In thin sections the iron-enriched parts

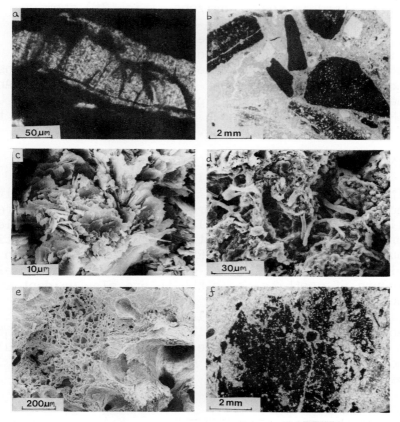

Fig. 2(a) A band of crystalline goethite in a
      laterised schist fragment;   (b) Laterised
      fragments of shales in a B horizon of an
      Ultisol;   (c) SEM micrograph of kaolinite flakes
      stacked vertically in the matrix of a laterised
      shale fragment;   (d)SEM micrograph of the matrix
      of a laterised shale fragment showing some
      tubular goethite in the voids;   (e) SEM micro-
      graph of voids in a laterised shale fragment
      showing a pitted appearance and framework
      structure: (f) Iron-enriched parts in the
      variegated horizon of an Ultisol showing no
      definite boundary in the s-matrix.

(plinthites) in these horizons (Fig. 2f) show no
definite boundary in the s-matrix. There are no
crystalline goethites or neo-ferrans. The petroplinthite,on the other hand,shows diffuse to well
defined irregular boundaries. Some void neo-ferrans
and patches of crystalline goethite are visible. Angular
quartz occurs as skeletal grains in some nodules. Under
SEM, kaolinite flakes have a similar morphology to that
in the laterised fragments.

MINERALOGY OF THE CLAY FRACTIONS AND NODULES
      The minerals in the clay fractions of all the
Ultisols are kaolinite, illite and quartz. Kaolinite
and quartz are dominant throughout the profile whereas
illite occurs only in the lower horizons, e.g. variegated
horizon of profile with petroplinthite. (The relative
abundance is judged from XRD peak intensities). In the
Oxisols, kaolinite is the dominant mineral and small
amounts of goethite and gibbsite are also evident in
the clay fractions.
      The well rounded nodules show a dominance of
kaolinite,with goethite, gibbsite and quartz present
in smaller amounts. In the irregularly shaped nodules
and platy laterised parent material, kaolinite and
goethite are dominant.

CONCLUSIONS
      The mineralogical and micromorphological study
indicates that lateritic soils in Malaysia can be
identified using the type of lateritic gravels as
one criterion. The roundness of the nodules, runiquartz and abundant crystalline goethite indicate that
they are allochthonous. The petroplinthite formed
in situ shows that iron oxides first segregate in the
lower horizon to form iron-enriched parts which could
later harden with the formation of goethite. Rock
fragments become laterised by iron oxides, which occur
along the schistocity plane, around voids and,in the
matrix,as droplets.

ACKNOWLEDGEMENTS
      The author would like to thank Universiti Pertanian

Malaysia for the financial support for this study, and
the laboratory assistants of the Soil Genesis Section
for the analyses and the preparation of thin sections.

REFERENCES
Brewer, R. 1976 .   Fabric and Mineral Analysis of Soils.
      Robert E. Kreiger Publ. Co., New York. 482 pp.
Comerma, J.A.,   Eswaran, H. and Schwertmann, U. 1977.
      A study of plinthite and ironstone from Venezuela.
      Proc. Conf. Classification and Management of
      Tropical Soils, Kuala Lumpur, Malaysia.
Eswaran, H. 1979. Micromorphology of Oxisols. In :
      F.H. Bienroth and S. Paramanathan (Ed), Proc.
      2nd Intern. Soil Classn. Workshop, Part 1,
      Malaysia, 61-74.
Eswaran, H., Sys, C. and Sousa, E.C. 1975. Plasma
      infusion - a pedological process of significance
      in the tropics. Ans. Edaf. y Agrobiol.,34,
      665-673.
Eswaran, H. and Daud, N. 1980.   A scanning electron
      microscopy evaluation of the fabric and mineralogy
      of some soils from Malaysia. Soil Sci. Soc. Amer. J.,
      44, 855-861.
Hamilton, R. 1964. A short note on droplet formation in
      ironcrusts. In : A. Jongerius (Ed), Soil
      Micromorphology. Elsevier, Amsterdam, 277-278.
Paramanathan, S. 1980. Petroplinthitic (lateritic)
      soils in Peninsular Malaysia. Notes for soil
      correlation tour of lateritic soils. Soil and
      Analytical Services Branch, Dept. Agric., Malaysia.
Stoops, G. 1970. Scanning electron microscopy applied
      to the study of a laterite. Pédologie, 22, 268-280.
Stoops, G. and Jongerius, A. 1975. Proposal for a
      micromorphological classification of soil materials.
      I. A classification of the related distributions of
      fine and coarse particles. Geoderma, 13, 189-199.

MICROMORPHOLOGY, MINERALOGY AND CHEMISTRY OF TWO BASALT-DERIVED SOILS IN THAILAND

M. Dabbakula Na Ayuduya, Chutatis Vanida and
Seesuwan Bongkoj

Micromorphology Section, Soil Analysis Division,
Department of Land Development, Bangkok, Thailand

ABSTRACT

Basalts with and without corundum occur scattered
throughout Thailand. Two soils, the Tha Mai series
of recognised high fertility on basalt with corundum,
and the Chok Chai series of low fertility on basalt
without corundum, have been compared from a micromorphological, mineralogical and chemical viewpoint.
The Tha Mai soil contained significant amounts of
collophane and other weatherable materials and a high
phosphorus content, whereas the Chok Chai soil
contained few weatherable minerals and was low in
phosphorus. The Tha Mai soil is classified as a Typic
Haplorthox and the Chok Chai soil as a Typic Haplustox.
Both soils contained some illuvial clay, though
generally less than 1%.

INTRODUCTION

The basalts of Thailand probably range from the
Tertiary to the Pleistocene in age. Although not
the main rock of the country, the soils derived from
basalts are particularly important from an agronomic and
horticultural standpoint. The soils vary widely in their
fertility. The fact that some have high phosphorus
contents is known but the reason for this is not
established.

The objective of the project summarised in this
paper has been firstly to determine whether the key
to differences in fertility between these soils can be
explained by their micromorphological and mineralogical
properties, and secondly,to examine their genesis with
special reference to the presence or absence of an
argillic horizon.

675

THE SOILS
     Two soils were sampled. One was of the Tha Mai
series developed on basalt with corundum from the
Chantaburi Province in eastern Thailand. The other
soil is of the Chok Chai series from the Nakhon
Ratchasima Province on basalt without corundum. The
Tha Mai soil is classified as Typic Haplorthox and the
Chok Chai soil as a Typic Haplustox (Beinroth and
Panichapong, 1979).
     The main analytical differences between the two
soils are summarised in Table 1. Further details
are available in Beinroth and Panichapong (loc. cit.)

Table 1 . Analyses of the two soils.

Tha Mai series

Depth (cm)	Horizon	Clay (%)	15 bar water (%)	Org. C (%)	pH (water)	CEC (me/ 100g)	Ext.iron (as Fe %)
0-20	A1	62	31	1.6	5.3	17.6	11
20-50	B21ox	53	31	0.8	5.3	11.2	9
50-95	B22ox	57	31	0.8	5.2	10.5	9
125-155	B23ox	70	33	0.4	5.2	8.9	10

Chok Chai series

Depth (cm)	Horizon	Clay (%)	15 bar water (%)	Org. C (%)	pH (water)	CEC (me/ 100g)	Ext.iron (as Fe %)
0-10/14	A1	51	18	1.4	4.4	8.2	3
10/14-36	B21t	62	19	0.6	4.2	5.8	3
60-86	B22t	63	19	0.3	3.8	4.9	3
120-156	B23t	63	18	0.1	3.9	4.7	4
185-220	B3	60	18	0.1	4.1	4.7	4

MINERALOGY
     There are important mineralogical differences
between the two soils. The Tha Mai soil has a light
mineral fraction consisting of 83% quartz and 16%
collophane compared with 99% quartz in the case of the
Chok Chai soil. The amounts of opaque minerals
(magnetite, ilmenite, hematite and limonite) are about
the same in the two soils (77-78%) but there are

significant differences beteen the two soils in other
heavy minerals. The Tha Mai soil contains 1.67% of
spinel + garnet, 0.14% zircon, 0.16% tourmaline and
19.4% weathered olivine. By contrast the other heavy
minerals in the Chok Chai soil consist of resistant
zircon (18.4%) and tourmaline (3.0%).
       The mineralogical analysis thus clearly demonstrates
that the Tha Mai soil contains more weatherable minerals
than the Chok Chai soil and a potential source of
phosphorus in the collophane.

CHEMISTRY
       The pH and base saturation of the Tha Mai soil
are higher than those of the Chok Chai soil except in the
case of the upper two horizons of the latter. The subsoil
base saturation falls to about 10 in the case of the Chok
Chai soil and to about 22 in the Tha Mai soil. The
CEC falls to 5 and 9 me/100g respectively.
       The most striking differences between the two soils
is in terms of available P and K as shown in Table 2.

Table 2 : Available P and K in the two soils.

	Tha Mai			Chok Chai	
Depth (cm)	Available P (Bray No.2) (ppm)	Available K (NH4Ac; pH7) (ppm)	Depth (cm)	Available P (Bray No.2) (ppm)	Available K (NH4Ac; pH7) (ppm)
0-20	55	86	0-10/14	5	5
20-50	55	74	10/14-36	2	23
50-95	65	43	60-86	2	12
125-155	95	12	120-156	1	12

MICROMORPHOLOGY
    Both soils lie within the Oxisol order and are
characterised by B horizons with subangular blocky
structure breaking down to micropeds. A summary of the
micromorphological properties of some horizons is given
below,using the terminology of Brewer (1964) and
Eswaran and Baños (1976):

Tha Mai 20-50 cm : Dark brown plasma; inundulic plasmic
fabric; low plasma birefringence; agglomeroplasmic
related distribution; peds separate into micro-peds;
common small iron oxide nodules; rare skeleton grains;
mainly silt-size quartz and few small pseudomorphic
altered pyroxenes; few root fragments.
    50-90 cm : As above but plasma is slightly lighter
coloured; slight increase in small grains of augite in
the groundmass; localised illuviation argillans, < 1%
overall.

Chok Chai 50 cm depth: Agglutinic specific related
distribution pattern; strong expression of aggregation
of plasma into micro aggregates; few fine sand-size
quartz; few fine sesquioxidic nodules; few patches of
strongly oriented illuviation argillans, overall < 1%.
    100 cm depth : Agglutinic specific related dis-
tribution pattern as above but more closely packed;
subcutanic orientation of plasma leading to ooidsepic
plasmic fabric; some patches of illuviation argillans;
some papules; some runiquartz; strong expression of
micropeds, perhaps associated with past activity of soil
fauna.

CONCLUSIONS
    The mineralogical analyses indicate that the Tha
Mai soil has a richer suite of minerals and a larger
content of weatherable minerals than the Chok Chai soil.
Hence,the potential of the soil to supply nutrients is
greater in the Tha Mai soil and this is reflected in
the greater fertility of this soil compared with the
Chok Chai.
    The high phosphorus content of the Tha Mai soil can

be linked at least in part to the presence of collophane.
    B horizons of both soils have well developed oxic
microfabrics although the amounts of weatherable minerals
in the Tha Mai soil are not typical of an oxic horizon.
There is some evidence for clay translocation in both
soils but amounts are less than 1%, although identification is difficult in the Chok Chai soil because of the
disrupted appearance. On the micromorphological evidence,
there would be some justification for examining further
Chok Chai profiles to determine whether they contain
argillic horizons.

ACKNOWLEDGEMENTS
    The authors thank the following: Soil Survey
field staff; the Cartographic section, the Printing
section; the Information and Extension section, Soil
Conservation Division, and Mr. J.F. Osborne.

REFERENCES
Beinroth, F.H. and Panichapong, S. 1979. Second Int.
    Soil Classification Workshop. Part II. Thailand.
    Soil Survey Division, Bangkok, 430 pp.
Brewer, R. 1964. Fabric and Mineral Analysis of Soils.
    John Wiley & Sons, New York, 470 pp.
Eswaran, H. and Baños, C. 1976. Related distribution
    patterns in soils and their significance. Anal.
    Edaf. Agrobiol., 35, 33-45.

THE EFFECT OF FIRE ON HUMUS PROFILES IN CANADIAN BOREAL FORESTS

B.T. Bunting

Pedology Laboratory, Department of Geography, McMaster University, Hamilton, Ontario, Canada

ABSTRACT
        The nature and extent of fire impact on boreal forest humus profiles is described. Recolonisation by moss, lichen and higher plants on sites of various age provide varied litters to supplement surviving humified material. New humus forms and modexi are dependent on intensity of past burns, partially-burned humus profiles showing more elaborate and varied components, intensely-burned profiles having many charred fragments of varied origin. The physical and chemical properties of the humus profiles of intensely-burned and partly-burned sites are significantly different, as are the proportions of skeletal, fibrous and plasmic materials.

INTRODUCTION
        The effects of forest fires on the humus profile (Babel, 1975) of boreal forest soils have rarely been investigated through the techniques of micromorphology, though numerous studies have been made of the effects of fire on forest nutrition (Wagle and Kitchen, 1971), and the forest-soils ecosystem (Ahlgren and Ahlgren, 1960; Vogt, 1970), especially in commercial forests with prescribed burning.
        Most forest fires in the northern ecotone are caused by lightning occurring along the Arctic front (Kershaw et al., 1975). Perhaps 60-70% of total land area is so burned and a high proportion of this is repeatedly burned at intervals of less than 100 years, hence most boreal forest humus profiles contain large amounts of charcoal. Within central and southern Canada, extensive wildland fires are caused by summer storm lightning, though many fires are caused by man (Donnelly and Harrington, 1978).
        Materials for routine analysis and micromor-

phological study were taken from northern Canada in the
Dubawnt watershed (62° N; 107° W) and sites in
central Ontario (45 to 50°N; 77 to 94°W). Samples
were taken from recently-burned humus profiles, before
and after the passage of fire. Samples from old-burn
forest floors were related to burn age as determined
by tree ring, fire scar or tree-core analysis. Soil
sampling provided basic information about fabric-related
properties to aid correlation of micromorphological
features and soil properties (Table 1).

METHODS
      Routine methods of analysis (McKeague et al.,
1976) were used in the physical and chemical analysis
of samples; the mineral samples were impregnated and
prepared by the method of Guertin and Bourbeau (1971);
and some fibrous samples prepared by using Paraplast
and microtome sectioning.

THE EFFECT OF BURN ON THE HUMUS PROFILE
      Most burns only affect the L and Fl horizons and
some woody components of the F2 horizon. The interplay
of many factors - moisture content, slope, wind turbulence
and density of tree cover - conditions the relative
impact of the burn. Humus profiles on dry elevated sites
may be entirely destroyed (Fig. la), especially on re-
burn; depression and lee sites may show no burn, but
contain added charcoal from fallout (after crown fires),
deposition from run-off, ash-blow and biologic transfer,
thus deepening the humus profile to 10 cm or more
(Fig. lb). Burn may alter many fabric-related properties
by dehydration, compaction, nutrient enrichment and
alteration of pH, CEC and acid components (Table 1),
thus changing the potential for biotic activity and
mineral weathering. In turn, recolonisation of burned
surfaces by plants is very different on partially and
intensely-burned sites.

Table 1. Fabric-related properties of burned humus profiles

	Depth (mm)	Fire event (yrsBS)+	Charcoal (%wt)	Ash (%wt)	Modexi (%area)*	Bulk density (g/cm³)	Porosity (%)	pH in CaCl₂ (0.01M)	CEC (meq)	Fibre (%)	C humic /C fulvic	Max. H₂O (%)
1. H horizon newburn	7	2	36	16.2	4	0.18	27	5.3	32	13.4	0.6	212
charcoal-free sample	..	..	<2	9.7	-	0.11	-	-	84	4.1	0.7	-
2. H Horizon older burn	22	32	26	8.2	12	0.16	23	4.7	71	10.7	0.4	298
3. Fl horizon recolonised burn	34	85	21	3.6	19	0.19	39	4.1	115	27.6	0.4	539
4. Fl horizon	30	124	13	4.0	24	0.21	35	4.3	101	25.0	0.6	374

* Percent area of thin section

+ Years before survey date (1976)

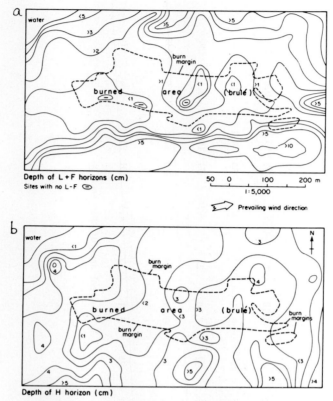

Fig. 1. Humus profile depths in an area of Dystric
         Brunisols showing, within the 'brulé', the thin
         litter and remaining humus under Cladonia lichen
         23 years after burn, and the uniform development
         of L & F and H horizons, 5-10 cm in thickness in
         unburned areas.

THE EFFECT OF INTENSIVE BURNING ON THE HUMUS PROFILE AND
ITS MICROMORPHOLOGY
     In the complete burn, roots are charred, the surface
ashified and mixed with burned fragments of woody debris
(Fig. 2a). This surface is covered by Polytrichum
poliferum moss and litter from shrubs (Ledum groen-
landicum) which form an F2 horizon within 10-30 years,
later to be covered by Cladonia lichen species after
50-80 years, in turn covered by a mat of Stereocaulan
lichen after 60 years or more (Kershaw et al., 1975).

Such initial L-Fl/L2F2 profiles in 'brulé' (burned
areas, Strang, 1973) are often only a few cm thick.
Where grazed by caribou, the lichen mat is < 1 cm depth.
Such organic profiles are compact, with many mineral
grains and charcoal nodules, ash-rich, of relatively
high pH, and subject to marked changes in thermal
regime (Rouse, in Kershaw et al. 1975). Intense fire
or prolonged natural summer heating leads to dehydration
and cracking of surface humus and grain coatings (Fig.
2b), the major skeleton being woody debris and nodules
of charcoal which, however, reveal a porous or
cellular aspect when dried and are not to be regarded as
typical β anthrocons (Bal, 1973).   Roots are rare,
some hyphae occur in mat-like arrangements, modexi are
degraded and rare, usually sited in nodule-free
enclaves.

THE CASE OF PROTECTED OR LESS-BURNED SITES
        Partially-burned F layers are more quickly
recolonised by mosses and vascular plants. The thicker
H horizons are more intensely reworked, in the manner
of a moder, often with an input of mossy debris in
the initial stages. However, the pre-fire laminar
arrangement of organic fragments is disrupted and
modexi develop only in areas of new or old unburned
F horizon material (Fig. 2c). Fresh needles are
bound together by hyphae in the L horizons and hardwood
fragments of large size are gradually incorporated in
the H horizons of older burns (> 120 years), otherwise
fibre content is low (Table 1). Some iron oxide,
released from mineral grains, coats organic fragments;
pollen grains are frequent and the proportion of
organic skeleton is low. Pore forms are more regular
but discontinuous and fine. Leaf-browning substances
are present in the southern sites to a considerable
degree, but are relatively scarce in the north.
        In both intensively-burned and partially-burned
forest humus profiles, the presence of charcoal nodules,
partly-burned fibrous debris, fungal hyphae and lichen
debris greatly affects the density and pore geometry
of the fabric (Table 1). Much of the unburned material

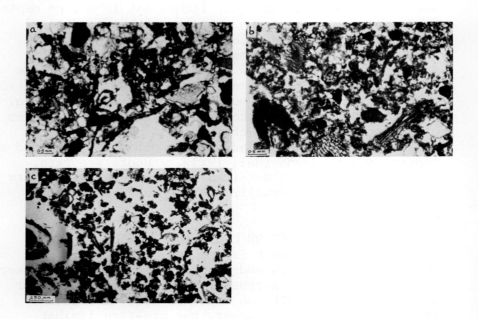

Fig. 2. (a) Charred root fragments and angular quartz
         in the Ah horizon of a recently burned site,
         a few modexi survive; (b) Dehydrated fibrous
         and woody debris, hyphae and charred fragments
         in an F2 layer, 55 years since burn; (c) Modexi
         developed in F2 layer, 120 years since burn.

is seen to be rapidly used by organisms and the development
of modexi is directly proportional to its amount.
    The moder type humus of the less-burned material is
only transformed into a more pitch-type amorphous F2-H
horizon after about 160-200 years of re-growth and this is
potentially very flammable, containing intercalated
needles and hyphae as well as resinous substances.

CONCLUSION
      Fire effects on humus profiles are varied and complex.
Though the major effect is of transformation to nodular
charcoal, this is mainly derived from woody debris and
its burning is often incomplete. Much of the organic
fabric in lower F horizons is only partially affected,
remaining fibrous or cellular, and is gradually altered
by recolonising organisms. Hyphae and new debris increase
in the early stages of redevelopment of dehydrated H
horizons. The history of plant recolonisation of the site
is the major influence on the post-burn humus profile.
Recognition of the proportion and type of burned organic
material, and of its removal before analysis, aids in
the correct assessment of the chemical properties of the
remaining unaffected humus, for the fabric-related and
chemical properties of burned and unburned humus in
any horizon are very different.

REFERENCES
Ahlgren, F. and Ahlgren, C.E. 1960. Ecological effects
      of forest fires. Botanical Review, 26, 483-533.
Babel, U. 1975. Micromorphology of soil organic matter.
      In : J.E. Gieseking (Ed), Soil Components. Vol. 1:
      Organic Components.   Springer-Verlag, New York,
      369-473.
Bal, L. 1973. Micromorphological Analysis of Soils.
      Soil Survey Papers No.6. Neth. Soil Surv. Inst.
      Wageningen, 174 pp.
Donnelly, R.E. and Harrington, J.B. 1978. Forest fire
      history maps of Ontario. Can. Forestry Service,
      Misc. Rept., FF-Y-6, Ottawa.
Guertin, R.K. and Bourbeau, G.A. 1971. Dry grinding of
      soil thin sections. Can. J. Soil Sci., 51, 243-
      248.
Kershaw, K.A., Rouse, W.R. and Bunting, B.T. 1975. The
      impact of fire on forest and tundra ecosystems.
      Dept. Indian and Northern Affairs, Arctic Land
      Use Res. Rep., 74-63, Ottawa, 1-81.
McKeague, J.A. (Ed), 1976. Manual on Soil Sampling
      and Methods of Analysis. Soil Research Inst.,
      Ottawa, Canada, 212 pp.
Strang, R.M. 1973. Succession in unburned subarctic

woodlands. Can. J. Forestry Res., 3, 140–143.

Vogt, R.J. 1970. Fire and northern Wisconsin pine barrens. Proc. Annual Tall Timbers Conf., 10, Fredericton, New Brunswick.

Wagle, R.F. and Kitchen, J.H., Jr. 1971. Influence of fire on soil nutrients in a Ponderosa pine type. Ecology, 53, 118–125.

MICROFABRICS OF A SPHAGNOFIBRIST AND RELATED CHANGES RESULTING FROM SOIL AMELIORATION

R.F. Hammond[1] and J.F. Collins[2]

[1] The Agricultural Institute, Peatland Experimental
Station, Lullymore, Co. Kildare, Ireland
[2] Dept. of Soil Science, Faculty of Agriculture,
University College Dublin, Dublin 4, Ireland

ABSTRACT
    The peatland resources of Ireland have been modified
by man for agriculture and by the production of hand-cut
fuel. The land resulting from the latter modification has
an economic potential for development. This study examines
the changes in microfabrics (O-fabrics) of soils formed
in raw peat materials after drainage and amelioration.
It has established that there are marked differences in
parent material O-fabrics and that techniques of amelioration have had a strong influence on surface horizon
microfabrics.
    The methodology of reclamation and time period over
which moulding has taken place correlates with the O-
fabrics described and with the physical and chemical
properties of the peat types described.

INTRODUCTION

    Peatland constitutes 17.2% of Ireland's land
resources, mostly as raised and blanket bogs. The raised
bogs occur mainly in the Midland Plain where they occupy
0.3 million ha (Hammond, 1979) and have over the centuries
been modified by man for agriculture and by the production
of hand-cut fuel (Fig. 1). The land type resulting from
the latter modification is very variable but has an
economic potential for land development. In recent years
small areas are gradually being developed for grass
production.
    The development of a productive fertile organic soil
from raw peat materials requires a considerable period of
time, being influenced by climatic, edaphic, anthropogenic
and other factors. To date, no studies have been
carried out under Irish conditions to quantify the changes
in organic soil fabrics (O-fabrics) resulting from reclama-

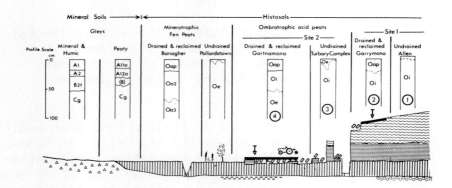

Fig. 1.   Schematic cross section of typical landscape
with organic soils in the Irish Midlands.

tion.   Documentary evidence of the early agricultural
development of peatlands is available, e.g. King (1685),
and the Bog Commissioners (1810-1814).   On the basis of
this information and present day knowledge of sites of
known reclamation history, a study was initiated
to document the changes in O-fabrics of peat materials
under Irish conditions.

SOILS AND THEIR SETTING
Two sites were selected on organic soils developed
in Sphagnum materials of similar properties and
characteristics but differing in age and method of
reclamation.   Site 1 had been reclaimed prior to 1870
(Fig. 2) and Site 2 in 1974.   The basic method of
reclamation at Site 1 [Bog Commissioners Reports (1810-
1814)] was to drain, apply calcareous marling materials
to the surface (880 t/ha loc. cit.) and incorporate by
digging.   In addition the bog surface may have been
pared, the parings burned and the ash residues incorporated.
Under modern reclamation techniques at Site 2 (Fig. 1)
ground limestone and nutrients are incorporated by rotary
cultivation following drainage and levelling.
Two soils were sampled at both sites;   one from the

1837   1870

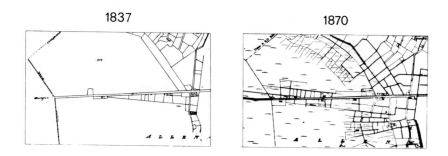

Fig. 2.   Maps taken from the initial Ordnance Survey
          of Ireland and subsequent revision showing the
          period of reclamation between 1840-1870 at
          Site 1.

reclaimed area and the second from an unreclaimed area
in close proximity. Profiles were sub-sampled at 10 cm
intervals and a range of physical and chemical
measurements (Lynn et al., 1974) made to characterise
the soils and show the changes occurring since initial
reclamation. The data presented in Table 1 are for
profiles 1,2,4 (Fig. 1). Profile 3 is not included
as the data are similar to that for Profile 1, Site 1.
At Site 2, samples for O-fabric analysis were taken
only from the surface 0-10 cm depth.
        Samples for micromorphological analysis were im-
pregnated according to the method of FitzPatrick and
Gudmundsen (1978) and thin sections prepared using
standard techniques. Thin sections were examined by
transmitted light using a polarising microscope.
Proportions of the components comprising the O-fabric
were calculated by counting features along random
transects at 16 μm intervals along bands 1.0 mm apart
to give 1,000 observations per thin section. The
components were separated mainly on the basis of
terminology used by Barratt (1968) with definitions
from Babel (1975).

Table 1. Physical and chemical data from unre-claimed (Pr1) and reclaimed (Pr2, Pr4) organic soils developed from ombrotrophic parent materials.

Depth cm	Peat Type* Designators			Moisture % dry weight			Ash % D.M.			Bulk Density g/cc			Rubbed Fibre % Vol.			pH (H$_2$O)		
	Pr1	Pr2	Pr4	Pr1	Pr2	Pr4	Pr1	Pr2	Pr4	Pr1	Pr2	Pr4	Pr1	Pr2	Pr4	Pr1	Pr2	Pr4
0-10	Nsd	Oap	Oap	897	318	440	4.1	27.9	13.3	0.143	0.268	0.208	12	2	8	3.72	6.40	6.92
10-20	hs	Oap	S	1076	486	614	0.7	11.8	3.4	0.094	0.259	0.148	6	2	26	3.61	6.20	4.15
20-30	Sca	hs	S	1117	646	860	2.2	7.2	2.3	0.064	0.132	0.081	24	10	44	3.62	5.58	3.59
30-40	hSer	Caer	Cascy	1316	754	809	1.2	2.4	2.0	0.064	0.089	0.095	48	8	14	3.81	4.90	3.62
40-50	S	Sca	Scy	1393	672	953	1.2	6.5	1.6	0.050	0.110	0.053	46	4	32	3.78	5.40	3.69
50-60	S	Cas	S	1692	747	997	1.4	8.6	1.7	0.061	0.105	0.094	56	4	26	3.81	4.78	3.89
60-70	Sca	Scyca	S	1872	793	1085	1.1	4.8	1.6	0.063	0.100	-	30	4	28	3.90	5.30	3.75
70-80	S	hs	S	1720	1073	1131	1.1	1.9	1.6	-	0.102	-	20	6	34	4.02	4.71	3.82
80-90	S	hs	S	2160	679	848	4.2	6.7	2.4	-	-	-	20	10	32	4.05	4.79	3.82
90-100	S	hs	Scacy	1753	897	826	1.3	1.6	2.3	-	-	-	50	4	36	4.08	5.50	4.32

* Nsd - Natural surface de-composition; h - humified; S,s, - Sphagnum; Ca,ca - Calluna;
er - Eriophorum; cy - cyperaceous; Oap - cultivated surface horizon

MICROFABRICS OF THE SOILS

The plant assemblages and their stratigraphy are given in Table 1 for the unreclaimed profile (Pr 1) and for both reclaimed profiles (Pr 2 and Pr 4). Taken together the profiles contained twelve different peat types to a depth of one metre. Consequently on re-clamation, the degree and nature of the processes involved in the transformation of organic parent materials will be strongly influenced by their initial state and each type will react differently to physical, chemical and biological ripening processes (van Heuveln et al., 1960). The data presented in Table 1 compare the physical and chemical characteristics of Profile 1 to reclaimed Profiles 2 and 4, which have undergone moulding (Pons, 1960). By comparing the data in Table 1, the changes taking place during the moulding can be quantita-tively assessed and can be related to the microfabrics (O-fabrics) observed in thin section. Fig. 3 illustrates the initial status of raw Sphagnum moss peat (parent material of the derived soils) and the changes and modifications within the organic matter profile during pedogenesis.

Fresh Sphagnum sampled from the undrained bog surface shows a very open Sphagno-humiskel fabric with

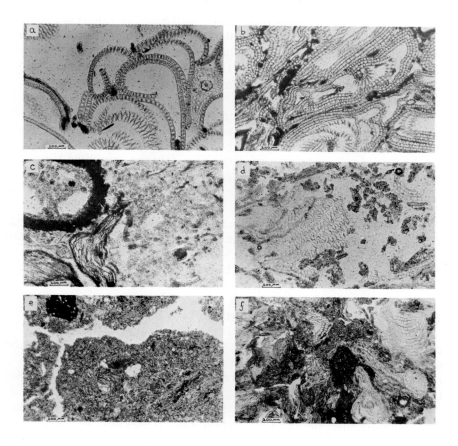

Fig. 3 (a) Very open Sphagno-humiskel O-fabric of
       fresh Sphagnum moss; (b) Open Sphagno-humiskel
       O-fabric with inclusions of humicol of poorly
     humified Sphagnum; (c) Plasmic humicol O-fabric
     with inclusions of lignic humiskel of humified
     Calluna/Sphagnum peat; (d) Moder O-fabric with
     disorientated weathered Sphagno-humiskel and
     humicol elements from the subsurface horizon;
     (e) Mullicol O-fabric with structural elements
     showing dense quartzitic humicol material from
     a cultivated and marled surface horizon under
     permanent pasture; (f) O- fabric from a
     cultivated surface horizon of a recently re-
     claimed organic soil sown to grass showing a
     random heterogeneous mixture of disrupted elements
     of Sphagno - humiskel and humicol materials.

no browning or formation of organic fine substance
(Babel, 1975) (Fig. 3a). Preserved Sphagnum mosses show
browning with an open Sphagno-humiskel fabric and a
layered preferred orientation. The voids between the
whorls of Sphagnum leaves are filled with humicol
materials (Fig. 3b).

O-fabrics of the Sphagno-humiskel type explain the
permeable nature and high water-holding capacity of
these materials. They also reflect accelerated peat
growth under wetter environments (Moore and Bellamy,
1974, p. 151). Intercalations of denser humicol
dominated O-fabrics with inclusions of lignic humiskel
components (Fig. 3c) are a common feature at various
levels in the upper tiers of a raised bog. Microfabrics
of this type illustrate the drier bog surface conditions
in the past (loc. cit. p.153) and/or humification-
decomposition processes within the bog during the post-
glacial period.

Irrespective of parent material microstructure,
lowering the water-table initiates moulding by
increasing biological activity and decomposition
processes. This increased activity is reflected in the
subsurface O-fabric from Profile 2 which contains
comminuted debris and weathered Sphagnum leaves typical
of moder O-fabric (Fig. 3d).

At Site 1, Profile 2, the added marling materials
and subsequent pedogenesis has had a marked effect
on microfabric structure in the surface horizon. The
humiskel components of the organic matter profile are
transformed to humicol materials and there has been an
intimate mixing with skeleton grains and plasmic
material of the added mineral fraction (Fig. 3e). This
moulding process has resulted in a mullicol O-fabric
with a stable fine structure under permanent pasture.

In comparison, after five years of reclamation
with lime and nutrients added, the O-fabric yields a
heterogenous mixture of weathered organic skeletal
materials and organic fine substance (Sphagno-humiskel
with humicol materials) (Fig. 3f) with various degrees
of browning. These have proceeded to the stage where
no intact plant organs other than recent roots are
discernible. This heterogeneity and lack of structure
is a reflection of the shorter time period in which
pedogenesis has proceeded and reflects a lower rate of

macro-biological activity in particular. Earthworms
occur more frequently in soils with mullicol O-fabrics,
as in the surface horizon of Site 2, than in soils
represented by Site 4, thereby reflecting the better
edaphic conditions of the former.

RELATIONSHIPS BETWEEN MICROFABRICS AND PHYSICAL AND
CHEMICAL DATA
        The O-fabrics described for the organic matter
profiles correlate with the physical and chemical data.
There is a direct relationship between bulk density
values and microfabric of the peat parent materials
under study. The denser humicol O-fabrics have bulk
density values from 0.10-0.11 g/cc and rubbed fibre
values of < 10%, while the open humiskel fabrics
range from 0.05-0.064 g/cc and rubbed fibre from 30-50%
respectively. The patterns of pH changes with depth
reflect the volume of liming materials added and
length of time available for incorporation and translocation. Despite gradual though erratic changes in
pH values to a depth of one metre in Profiles 2 and 4,
the sharp breaks at the base of the Oap horizons in
both are noteworthy.
        The differences in the parent material O-fabrics can
also be correlated with the moisture values. Moisture
values of 1400-2100% approx. are associated with a
larger volume of inter- and intra-cellular spaces in
humiskel fabrics, while values of 700-1100% approx.
are more typical of denser humicol fabrics both in
drained and undrained sites.
        The addition of liming materials to the soil surfaces
of Profiles 2 and 4 has increased ash contents 10-20
fold and bulk density values to greater than 0.2 g/cc.
The ash values to 70 cm in Profile 2 exhibit the effects
of greater volume of material added to the surface and
degree of incorporation and translocation. The result
of this treatment is especially noticeable in the
surface 30 cm where values range from 7-27% ash. Such
additions have had a material effect on soil development
as illustrated by the microfabrics.
        The surface horizon of Profile 2 exhibits an O-fabric
with well developed fine structural units having sharp
boundaries and accordant ped faces. The individual
aggregates ($\emptyset \approx 1mm$) consist of mineral grains embedded

in dense organic plasma. This type of mullicol O-fabric
correlates with bulk densities > 0.25 g/cc, rubbed
fibre values of ≤ 2.0%, ash values > 11.0% and the
homogenisation effect of the moulding processes.

In contrast the heterogeneous O-fabric from the
surface horizon of Profile 4 is dominated by random
humiskel and humicol elements. This is caused mainly
by mechanical soil amelioration rather than biological
activity. The addition of liming material is reflected
in increased bulk density and ash values while the
decrease in rubbed fibre can be attributed to increased
decomposition as a consequence of soil amelioration.

The soil materials studied have illustrated the
different microfabrics associated with peat parent
materials and the changes and regroupings taking place
with time in response to divergent methods of soil
amelioration. It is of interest to establish how long
it takes to form an homogenised mullicol O-fabric. To
this end, the study has been extended to other locations
which have undergone similar treatments at intermediary
times since the 1850s.

ACKNOWLEDGEMENTS
The technical assistance of Mr. E. Brennan is
gratefully appreciated in the preparation of these
sections and carrying out laboratory studies.

REFERENCES
Babel, U. 1975. Micromorphology of soil organic matter.
    In: J.E. Gieseking (Ed), Soil Components. Volume 1:
    Organic Components.Springer-Verlag, New York,
    369-473.
Barratt, B. 1968. A revised classification and nomen-
    clature of microscopic soil materials with particular
    reference to organic components. Geoderma, 2,
    257-271.
Bog Commissioners (1810-1814). Bogs in Ireland the
    Practicability of Draining and Cultivating them.
    Vols. I-III.
FitzPatrick, E.A. and Gudmundsson, T. 1978. The impreg-
    nation of wet peat for the production of thin
    sections. J. Soil Sci., 29, 585-587.
Hammond, R.F. 1979. Peatlands of Ireland. Soil Survey
    Bulletin No.35. An Foras Taluntais, Dublin.

King, W. 1685.   On the bogs and loughs of Ireland.
     Phil. Trans., 15, 948.
Lynn, W.C., McKinzie, W.E. and Grossman, R.B. 1974.
     Field laboratory tests for characterisation of
     Histosols.   In: A.R. Aandahl. No.6 S.S.S.A.
     Special Publication Series, Madison, Wisc., 11-20.
Moore, P.D. and Bellamy, D.J. 1974.   Peatlands.   Elek
     Science, London, 221 pp.
Pons, L.J. 1960.   Soil genesis and classification of
     reclaimed peat soils in connection with initial
     sod formation.   7th Intern. Congr. Soil Sci.,
     Madison, Wisc., IV, 205-211.
Van Heuveln, B., Jongerius, A. and Pons, L.J. 1960.
     Soil formation in organic soils.   7th Intern.
     Congr. Soil Sci., Madison, Wisc., IV, 195-204.

MICROMORPHOLOGY OF AN ORTHIC TURBIC CRYOSOL - A PERMAFROST SOIL

C.A. Fox

Land Resource Research Institute, Agriculture
Canada, Ottawa (LRRI Contribution No.116)

ABSTRACT
      The frozen layer of the Orthic Turbic Cryosol was
characterised by distinct fabrics, conglomeric and
orbiculic arrangements, indicating displacement of the
f-members and finer f-matrix. Cryogenic processes
were found to have a major effect on the soil fabric.

INTRODUCTION
      The Orthic Turbic Cryosol is a mineral soil
generally associated with patterned ground features;
it has permafrost within 2m of the surface and shows
marked evidence of cryoturbation (Canada Soil Survey
Committee, 1978). Its appropriate equivalents in the
USA and FAO taxonomies are respectively Pergelic
Cryaquept and Gelic Cambisol. Because of the relatively
recent introduction in 1976 of the Turbic Cryosol Order
to the Canadian system of soil classification, very little
was understood regarding the relationship between microfabrics and the cryogenic processes affecting the pedon.
Preliminary observations of the microfabrics of some
permafrost soils associated with earth hummocks were
reported by Pettapiece (1974) and Brewer and Pawluk
(1975). They observed changes of microstructure with
depth. However, the observations were limited to the
active layer and there was no emphasis on the characterisation of the fabrics which occurred either within or
in close proximity to the frozen zone. Therefore, the
purpose of this work was to examine thin sections of an
Orthic Turbic Cryosol pedon in order to determine
whether cryogenic processes have caused distinct fabric
arrangements to occur within the frozen layer.

SITE DESCRIPTION AND METHODOLOGY
      The site (65° 10' N Lat., 127° 27' W Long.) is

situated at an elevation of 244 m (a.s.l.), mid-slope
on a rolling till plain (slopes 5-9%) in the Mackenzie
Plain near the Carcajou River of the Mackenzie river
valley. The surrounding vegetation is open black
spruce-lichen-moss woodland of the Subarctic Boreal
Forest region. At the time of sampling (Fig. 1) the
frost table was at 25 cm and the rooting zone was 20
cm. The pedon (Y39), an imperfectly drained earth
hummock and the associated interhummock, was selected
from ten Orthic Turbic Cryosols that had been
examined previously by Fox (1979). The micromorphology
was described according to terminology of Brewer (1964),
Brewer and Pawluk (1975) and Fox (1979). Qualitative
assessment of the relative occurrence of specific
morphology was according to the following: VF - very
frequent; F - frequent; C - common; O - occasional;
R - rare (very hard to locate).

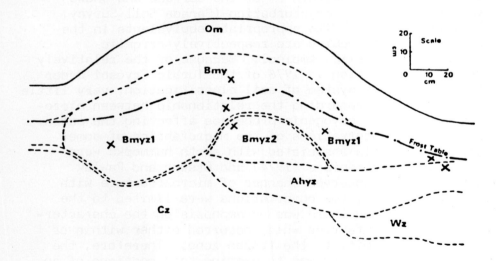

Fig. 1

MICROMORPHOLOGY
     The micromorphological description is as follows:

Bmy (0-25 cm)
     The horizon is dominated by dark greyish brown
granular (400-2000 μm diam.) structure. Compound packing
voids and craze planes result (Fig. 2a). Voids vary from
100 μm to > 2mm with some of the larger pores occupied
by roots. Numerous charred organic fragments were
observed with visible cell structure. Skeleton grains
dominantly less than 100 μm are often surrounded by
yellowish orange embedded grain ferri-argillans < 10 μm
thick.
(x25) phyto-matrigranic = phyto-matrigranoidic
(x100) skelsepic porphyroskelic

Bmyz1 (25-38 cm)
     In some regions, matrigranic f-members (40-800 μm)
and organic glaebules (600-1400 μm) are enclosed (Fig. 2b)
by the finer matrix material to form a conglomeric
porphyroskelic distribution. (This term is discussed in
Fox and Protz (1981)). The subangular to angular skeleton
grains (<16-300μm) tend to occur in circular arrangements
(orbiculic fabric) (Fig.2c). Charred organic fragments
(< 1mm) are a dominant component of the f-matrix. Skew
planes (80-200 μm wide) are (C-F); irregular, prolate,
and circular vughs (200-800 μm, few 1.4 to 2.0 mm long)
are (0-C); and diagonal trending channels (200-800 μm
wide) are (C). Rare ferri-argillan (60-80 μm wide)
occurs in a vugh (Fig. 2d) or may also occupy former
channels. Embedded grain ferri-argillans (C-F) (< 5-20
μm) are discontinuous to continuous around the skeleton
grains and become more intense in the mottled regions.
(x25) matrifragmoidic/conglomeric-porphyroskelic/
porphyroskelic
(x100) latti-skel-insepic with orbiculic porphyroskelic

     Fig. 1. Diagrammatic cross-section of Y39 showing
          sampling points (adapted from field notes by
          C. Tarnocai).

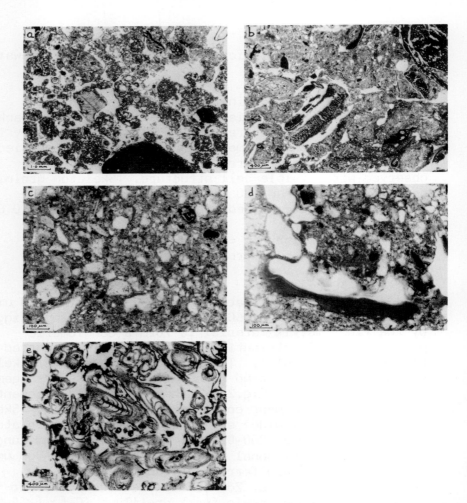

Fig. 2(a)   Matrigranic and matrigranoidic distribution
in Bmy horizon. PPL;   (b) Granic f-members
enclosed into f-matrix material forming conglomeric
porphyroskelic fabric distribution, Bmyzl. PPL;
(c) Skeleton grains distributed into circular
pattern (orbiculic porphyroskelic fabric).   PPL;
(d) Ferri-argillan (80 μm wide) deposited in void,
Bmyzl. PPL;   (e) Humi-phytogranic distribution of
Ahyz. PPL.

Bmyz2 (38-57 cm)
     The morphology of this horizon is very similar to
Bmyz1. Larger f-members (1.6 to 6 mm) are separated by
diagonal downward trending channels (80-400 μm wide).
Internally, the f-members occasionally show a conglomeric-
porphyroskelic fabric with fairly distinct embedded
granular units (300-600 μm).
(x25) conglomeric porphyroskelic/matrifragmic/matri-
fragmoidic/porphyroskelic
(x100) latti-skelsepic with orbiculic porphyroskelic

Ahyz (57-79 cm)
     A very porous f-matrix of intact moss fragments
(400-2000 μm) and humified f-members (60-300 μm) (Fig.
2e). Mineral grains are absent. Rarely occurring
large (1.2 to 2.0 mm) elliptical to rounded faecal
pellets are vertically aligned. Smaller faecal pellets
(<300um) are randomly distributed (R-0) within the
f-matrix.
(x25) humi-phytogranic--humi-phytogranoidic

RESULTS AND DISCUSSION
     The fabric in the Orthic Turbic Cryosol pedon
varies with depth from a granular distribution in the
Bmy horizon to mainly porphyroskelic fabric with
fragmic and fragmoidic components in the Bmyz2 to distinct
organic f-members in the Ahyz. In the Bmyz1 and Bmyz2
horizons, both located within the frozen layer, conglomeric porphyroskelic and orbiculic porphyroskelic fabrics
were observed. No thin sections were available for the
C horizon of this pedon, but the C horizons of the other
Orthic Turbic Cryosol pedons developed in similar parent
materials (clay loam) were studied. Their fabrics were
porphyroskelic, but in the vicinity of the Ahyz horizon
the fabric was dominated by finely disseminated organic
material and showed evidence of displacement of the
skeleton grains. Observations of orbiculic and
conglomeric porphyroskelic fabric in the frozen layer
near the vicinity of the frost table suggest that
cryoturbation processes have led to the displacement of
the f-members (Fox and Protz, 1981).

Cryostatic pressures may also account for the intrusions of the organic f-members of the Ahyz into the material Bmyz1 and Bmyz2 horizon and along the interface between the Bmyz1 and Cz horizon.   Zoltai et al. (1978) reported for pedon Y39 that younger organic materials (880±80) occurred together with older material (1020±80) in the central portion of the earth hummock suggesting that intrusion and displacement of material had taken place.

In the Bmyz1 horizon, within the frozen layer, argillans were observed in some of the voids and channels indicating that during the thaw cycle substantial suspended material may be present in the soil water.   As water is removed for ice lens formation, the suspended material may become deposited on the void walls. Freezing and thawing cycles may also account for the development of the skelsepic fabric component in the soil matrix. For instance, cryostatic pressures may lead to orientation of the clay material at grain surfaces.

CONCLUSIONS

The fabric of the Orthic Turbic Cryosol pedon especially in the frozen layer is dominated by evidence of cryogenic processes. Conglomeric and orbiculic fabric components indicate that displacement of the f-members within the f-matrix has occurred. Such fabrics could be designated as sorted fabrics to characterise the distinct morphology that results when the f-members undergo reorganisation in relationship to the f-matrix.

ACKNOWLEDGEMENTS

The soils were sampled by C. Tarnocai (Manitoba Soil Survey 1973); thin sections prepared by R. Guertin (Land Resource Research Institute (LRRI) Ottawa) and lent for the study by J.A. McKeague (LRRI, Ottawa); funding was assisted by R. Protz (University of Guelph) and a National Research Council of Canada Scholarship.

REFERENCES
Brewer, R. 1964. Fabric and Mineral Analysis of Soils.
    John Wiley & Sons, New York, 470 pp.

Brewer, R. and Pawluk, S. 1975.   Investigations of some
     soils developed in hummocks of the Canadian Sub-
     Arctic and Southern Arctic regions.   1. Morphology
     and micromorphology.   Can. J. Soil Sci., 55,
     301-319.
Canada Soil Survey Committee, 1978.   The Canadian
     system of soil classification.   Can. Dept. Agric.
     Publ. 1646.   Supply and Services Canada, Ottawa,
     Ontario, 164 pp.
Fox, C.A. 1979.   The soil micromorphology and genesis of
     the Turbic Cryosols from the Mackenzie River Valley
     and Yukon Coastal Plain.   Ph.D. Thesis, Dept. Land
     Resource Science, Univ. Guelph, Ont. 196 pp.
Fox, C.A. and Protz, R. 1981.   Definition of fabric
     distributions to characterize the rearrangement
     of soil particles in the Turbic Cryosols.   Can. J.
     Soil Sci., 61, 29-34.
Pettapiece, W.W. 1974.   A hummocky permafrost soil from
     the Subarctic of Northwestern Canada and some
     influence of fire.   Can. J. Soil Sci., 54, 343-355.
Zoltai, S.C., Tarnocai, C. and Pettapiece, W.W. 1978.
     Age of cryoturbated organic materials in earth
     hummocks from the Canadian Arctic.   In: Proc.
     3rd Int. Conf. Permafrost (July 10-13, 1978),
     Edmonton, Alberta. Volume 1. Nat. Res. Council
     Canada, 325-331.

Author Index